MOONSHOT
MOMENTS

USHERING IN THE NEXT HUMAN RENAISSANCE
THROUGH AI, TRANSHUMANISM, AND PSYCHEDELICS

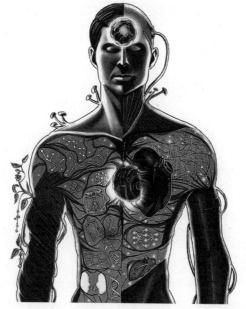

MILAN KORDESTANI

Health Communications, Inc.
Boca Raton, Florida

www.hcibooks.com

Library of Congress Cataloging-in-Publication Data
is available through the Library of Congress

© 2025 Milan Kordestani

ISBN-13: 978-07573-2524-3 (Paperback)
ISBN-10: 07573-2524-6 (Paperback)
ISBN-13: 978-07573-2525-0 (ePub)
ISBN-10: 07573-2525-4 (ePub)

Publisher: Health Communications, Inc.
        301 Crawford Boulevard, Suite 200
        Boca Raton, FL 33432-3762

Cover design by Ankord Media
Interior design and formatting by Larissa Hise Henoch

CONTENTS

Part III: Preparing the Self for Greater Creativity

Part IV: Embracing Our Cosmic Destiny

FOREWORD

Every great leap in human history begins with an idea that seems impossible to the people of that time. From Kennedy's declaration to reach the moon, to the emergence of human civilization in our evolutionary past, transformative change often appears to defy our understanding of how progress typically occurs, which is through a relatively slow and incremental process. But what if these seemingly miraculous leaps reflect a deep cyclical pattern in nature—one that could help us navigate our own transformative moment in history?

The term "moonshot moment," from which this book gets its title, traditionally refers to ambitious, groundbreaking projects that address huge problems through radical solutions and breakthrough technology. Google and similarly powerful organizations use it to describe initiatives that aim to solve massive challenges through revolutionary innovations. But the examples described in this book suggest something more fundamental: these moments reflect a universal pattern of transformation that occurs not just in human endeavors, but in the evolutionary process itself.

Consider what anthropologists call the "cognitive revolution"—a period between 100,000 and 40,000 years ago when our ancestors suddenly developed all the complexities that make us "human," like abstract thinking, symbolic language, and a new level of self-awareness. This rapid transformation, occurring in what amounts to an evolutionary blink of an eye, challenges our traditional understanding of how adaptive change occurs. It points to a deeper pattern present in cosmological, biological, social, cultural, and technological evolution: the *phase transition*. From the emergence of atoms to the birth of cells, from the rise of consciousness to the development of culture, reality organizes itself through a nested series of phase transitions, where smaller systems unite to form larger ones with new emergent properties.

If this philosophy is correct, then Milan Kordestani's moonshot moments aren't random occurrences; they are, to some useful degree, *predictable* and follow mathematical and dynamical principles we're beginning to understand. Some of these principles relate to thermodynamics, information, and computation, which can be unified under a new theory of evolution, what I have called "the Integrated Evolutionary Synthesis" in my book *The Romance of Reality*, which understands life and all its cultural products as an attempt by organized knowledge to survive against the natural tendency toward disorder famously associated with the second law of thermodynamics. This thermodynamic framing illuminates the computational and goal-directed nature of life, and reconceptualizes the story of nature as a game where world-modeling agents develop increasingly sophisticated strategies to survive in an unpredictable universe.

Each major evolutionary transition represents a leap in both functional organization and error-correction capability. These 'errors' aren't just mistakes—they're threats to the stability of the computational system, which could be an individual like a human, or a society. For instance, the emergence of language allowed humans to share knowledge and correct misunderstandings more effectively. Similarly, the scientific method represented another leap, providing systematic tools for discovering truth and correcting errors in our collective understanding. Each transition creates new ways for systems to maintain themselves against chaos, often through increasingly sophisticated organizational structures. The division of labor enables specialization, while the hierarchical modularity of these systems contains errors and prevents system-wide collapse. Life's multi-level architecture reveals itself as a design optimized for resilience, robustness, and adaptability.

Today, we stand at the threshold of another such transition. The emergence of artificial intelligence, particularly large language models, represents just one aspect of a broader phase transition in human civilization. We're witnessing the potential emergence of what Teilhard de Chardin envisioned a century ago as the "noosphere"—a global network of human computation and consciousness that represents a new level of planetary organization. Think of technologically aided humanity as forming a "global brain," creating a cognitive layer that integrates with and enhances the biosphere.

The culmination of this transition will be a "singularity" event, but not like many have imagined. The singularity is not a point at which AI surpasses human intelligence and makes us obsolete, nor is it a point where we evolve into machines ourselves by replacing our biology with technology. We will indeed integrate with our technology—in fact, we already have just by owning a smartphone —but it will be a different kind of event: one where we become in harmony not just with our technological products, but with the living planet as a whole. It is the moment when we recognize that we are the self-aware layer of the planetary superorganism, which has been called "Gaia" by the poetic naturalists and spiritual scientists who see the living world as one interconnected entity; not fully distinct from the inanimate natural world, but an agentic outgrowth of it.

In alignment with this view, Kordestani's book explores moonshot moments not just as historical events but as active templates for understanding and guiding transformation. These patterns help us understand not just what changed in the past, but how change itself works. They offer practical guidance for steering our own moonshot moments toward positive outcomes.

The book explores not just isolated moonshot moments, but "moonshot clusters"—periods where multiple transformative changes converge to create social and cultural phase transitions. Beyond merely chronicling these historical leaps, it invites readers to become active participants in the next great transition. Rather than passive observers of change, we are called to become conscious agents of transformation, helping shape the future we want to see.

The coming transition involves multiple catalysts that Kordestani recognizes: artificial intelligence provides new capabilities for processing and organizing information; transhumanist technologies suggest ways to enhance human capabilities; and psychedelics offer tools for shifting perspective and seeing the interconnected nature of reality. These aren't separate developments but aspects of a single metasystem transition in human civilization.

Understanding these evolutionary-developmental patterns is crucial as humanity faces unprecedented global challenges. Climate change, infectious disease, economic inequality, war and social unrest—these challenges can't be solved through gradual improvement alone. They require the kind of fundamental reorganization that characterizes true moonshot moments, and such transformations demand the participation of everyone. Think of the present transition as being analogous to the emergence of the first civilization, but now our civilization is about to become truly planetary, as a singular coherent and harmonious computational entity. We are all co-creators in the *Game of Life,* a "reality game" that has a narrative that is partially predetermined and partially 'open,' waiting to be *determined by us.* We, as free agents with the ability to reason counterfactually and choose the future path that is consistent with our goals and vision, stand to write the most epic chapter of history so far in the story of cosmic evolution.

This understanding transforms us from passive observers into conscious participants in humanity's next great leap forward. As our civilization approaches its own moonshot moment, facing challenges that seem to require impossible leaps, these recurring patterns offer both hope and direction. They remind us that throughout history, seemingly insurmountable barriers have been overcome through dramatic reorganization efforts by the collective that transformed what was realistically possible.

Kordestani's proposal for a personal knowledge management (PKM) system aligns with this view of evolutionary transitions as leaps toward new regimes characterized by new worldviews that bring about revolutions in understanding. Just as previous transitions involved new ways of organizing information and understanding reality, the current transition requires new tools for managing knowledge and navigating complexity.

The major themes in this book—AI, biohacking, hyper-cooperation, and the exploration of consciousness via mind-altering experiences—come together under the umbrella of moonshot moments. They represent different aspects of humanity's innate desire to push the bounds of what we think is possible and aim for the seemingly unachievable. This is all done not only in an effort to further humanity, but also to satisfy our curiosity and our natural urge to understand the world around us. It is all one project. It is a part of the process of natural uncertainty reduction and sense-making activity that we've had to do for as long as we've existed, in order to resist entropic decay.

This book serves as an instruction manual for the modern human, the "metamodern" human, and for our civilization as a whole. It's a guide to transforming the Game of Life into a truly "infinite game," ensuring the continuation and evolution of this grand project we call life. By understanding the dynamical patterns of transformation that bring new levels of order, complexity, and awareness into existence, we become better equipped to navigate and guide the changes ahead. The future isn't something that simply happens to us; it's something we help create through our ability to apply our knowledge with ideas, actions, plans, and policy. This book is an invitation to join the collective game that is the puzzle nature is counting on us to solve, as we are the vehicles through which reality becomes aware of itself. If we can become "meta-aware," or aware of our role in the grand evolutionary process, we can all meaningfully contribute to the greatest story ever told—the story of reality waking up. The most interesting part of the story is that it is doing so *through us*. What will be the moonshot moment you help to catalyze?

—**Bobby Azarian, PhD,** author of *The Romance of Reality*

PREFACE

A Quiet Dread . . .

For a while, a nagging unease had settled in my gut.

My previous book tackled the delicate dance of human discourse, the art of weaving empathy and common ground into divisive conversations. It was an honoring of the untidy, lovely way we are all connected. But as I finished that writing process, a different conversation began to drown out the echoes of the last. This one was about automation and robots, about the future of work, and the very essence of human existence.

Perhaps, like me, you've found yourself nervously browsing through a news-feed bursting at the seams with advances in artificial intelligence. Each story promises a future in which machines handle tedious tasks, freeing us to pursue . . . well, what exactly? As a species, we've always found meaning and purpose in the work we do, the challenges we overcome. What happens when the machines take over the mundane, leaving us adrift in a sea of leisure? This disquietness, this yearning for a future in which technology elevates our humanity, not diminishes it, is the seed from which this book sprouted.

Driven by a burning curiosity, I embarked on a research odyssey. I devoured the works of prominent artificial intelligence theoreticians like Ray Kurzweil and Max Tegmark and tech developers like Mustafa Suleyman. Each offered glimpses into a future shaped by intelligent machines. Every podcast featuring purpose-driven AI architects became my soundtrack, each voice a brushstroke on the canvas of this technological revolution.

The lines between theory and experience soon began to blur. My own mental and physical health journeys, with cutting-edge treatments like psychedelic therapy and prescribed peptide therapy for an injury, suddenly placed me at the crossroads of biohacking and transhumanism—two arenas that I believe offer great potential for our species, but that, like AI, are often met with fear and skepticism. The abstract debates on AI ethics suddenly intertwined with the very real possibilities of human augmentation.

Then came the revelation. Michael Pollan's captivating Netflix documentary series, *How to Change Your Mind*, introduced me to the transformative power of psychedelics. I found Pollan's investigation into the possibility of using psychedelics to open up fresh perspectives to be profound, particularly in the context of future technologies. Could an expansion of human consciousness and empathy to meet the evolving landscape be the key to navigating a future brimming with technological advancements? After watching the documentary, I read Pollan's companion book in its entirety, as well as the writings of Ben Sessa, Rick Doblin, Paul Stamets, and other authors, in an attempt to learn more about the science underlying these drugs of mass deception and their possible influence on human experience.

Inspiration struck from unexpected corners as well. My father is a technologist who also enjoys biohacking and occasionally philosophy. He suggested that I read the writings of Buddhist philosopher Alan Watts, which led me to the work of Aldous Huxley. Both writers' ability to unite traditional wisdom with contemporary ideas opened my eyes to new ideas about the possibilities of humanity and our role in the cosmos. I continued to discuss the looming questions around my research with my friend and colleague Chloe, a die-hard science fiction aficionado. She urged me to explore the worlds of Neal Stephenson and the books and films of Phillip K. Dick. These fictional imaginings, where humanity grappled with the consequences of technological advancements, provided fertile ground for thought experiments. There, I could grapple with the possibilities and pitfalls of a future intertwined with new technologies, all from the safety of my couch.

While I was doing an unconnected interview with a bright theoretical mathematician, our discussion evolved into an unanticipated digression into the difficulties of human collaboration, which ultimately had a vital role in influencing my research. Dr. Benjamin Kuipers's enthusiasm for Robert Wright's work on hyper-cooperation piqued my interest. Here, it seemed, was another crucial piece of the puzzle—hyper-cooperation was a framework for understanding how humans could not only survive but thrive in the face of transformative technologies.

All these influences became a whirlwind of notes, ideas, and arguments. Though each piece struck a deep chord, they all stayed stubbornly apart, a disorganized mass that yearned for a cohesive story.

But one recurring theme kept jumping out—moonshot moments. These weren't just incremental improvements or technological tweaks; these were bold leaps, moments of extraordinary, wild, and purposeful thinking. I saw them everywhere—in the books I devoured on transhumanism, the videos exploring AI trends, even in the philosophical treatises on altered consciousness. These weren't just abstract concepts. They were woven into the very fabric of these discussions, highlighting humanity's relentless pursuit of the seemingly impossible and capacity for unfathomable creativity.

AI, transhumanism, psychedelics, hyper-cooperation, the exploration of consciousness, even the anxieties about a future powered by machines all started to come together under the umbrella of moonshot moments. This book evolved to focus on the innate human desire to push the bounds of what we think is possible and aim for the seemingly unachievable.

We'll walk through the historical instances of moonshot moments that have redefined our world, and explore the potential paradigm shifts brewing on the horizon. But, rather than merely applauding ambition, we'll get into the moral issues, possible dangers, and essential actions to guarantee that these breakthroughs are beneficial to all people.

... Eased by Hope

While this book evolved mightily over the course of its life, it began with the goal of bridging the gap between generations when it comes to artificial intelligence (AI). Older generations, often frightened by dystopian science fiction and the rapid pace of technological change, felt a deep techno-skepticism. For them, AI was a Pandora's box best left unopened. I'd hoped to play the role of a friendly ambassador, dispelling myths and highlighting the incredible potential of AI for good. However, as I dug deeper into my research, a new and unexpected narrative emerged.

While older generations worried about the dangers of overoptimism, I started noticing a different kind of fear manifesting among younger demographics —a sense of overwhelming hopelessness. Gen Z and Generation Alpha, bombarded with news cycles dominated by climate change, political turmoil, and social strife, often felt like passengers hurtling toward an uncertain future on a runaway train. This, coupled with a constant barrage of information and the pressure to curate a perfect online persona, has resulted in a significant rise in mental health crises among our younger generations. I have not found myself immune to such pressures.

This was enlightening. Though many in my generation feel completely helpless to influence the course of technology, they are inheriting a world full with possibilities. I started to wonder: how can we, in this future we're going to inherit, be empowered rather than paralyzed? AI and a plethora of other emerging technologies offer an unprecedented opportunity: either choose to accept a dystopian future; or cooperate responsibly, with an open mind about the opportunities and challenges we face, and work together to build a better tomorrow.

This realization reshaped the trajectory of this book again. Reigniting a sense of possibility of hope, as a partial attempt to address the growing mental health crises of our time, became my mission. This meant a deeper exploration of various ideas that can aid that suffering—from the role AI can play in returning to us the gift of time to transhumanism's ability to improve our quality of life; to the growing mentalscapes possible through synthetic biology, psychedelics, and more.

This book is not a utopian fantasy, nor is it a blind endorsement of technological progress without concern for the very large ethical questions at play. It's a call to action to do all we can to explore and promote creating moonshot moments, those eras of stunning advancement and progress that have continually propelled our species into better times.

We can reject apathy, complacency, and fear and embrace a proactive, measured vision for the future.

Make no mistake, this will require effort. It will necessitate critical thinking, open dialogue, and a willingness to grapple with the complex ethical questions that arise as technology advances. It will require embracing powerful win-win mindsets in an era plagued with fears of scarcity. It will require, most of all, wildly bold thinking that produces moonshot moments and golden ages of progress and innovation. But the alternative—a future defined by fear, resignation, and a lack of purpose—is simply unacceptable.

This book is also an invitation—to younger generations especially—to shed the mantle of hopelessness and step into the driver's seat. It's an invitation to understand AI as a powerful tool waiting to be shaped. It's an invitation to become a better version of yourself and make moonshot moments of your own. It's an invitation to embrace your agency, to engage with the society you will inherit, and to champion a future that prioritizes progress and hope.

The pages that follow are meant to serve as a starting point for polite discussion and dialogue. After all, cooperation is essential to a better future filled with incredible opportunities. You'll hear about possible hazards as well as innovative success stories. You'll be given the opportunity in Chapter Three to choose *your* future and how you will contribute to our species as we make better choices about our destiny.

The ultimate goal is to enable you, the reader, to take a proactive role in determining the course of our shared future.

We, as a species, have faced existential challenges before. We tamed fire and lived to tell the tale. This time, however, we have the potential to craft a cosmic future brimming with possibility. The choice, ultimately, is ours.

With hope and anticipation,
Milan Kordestani

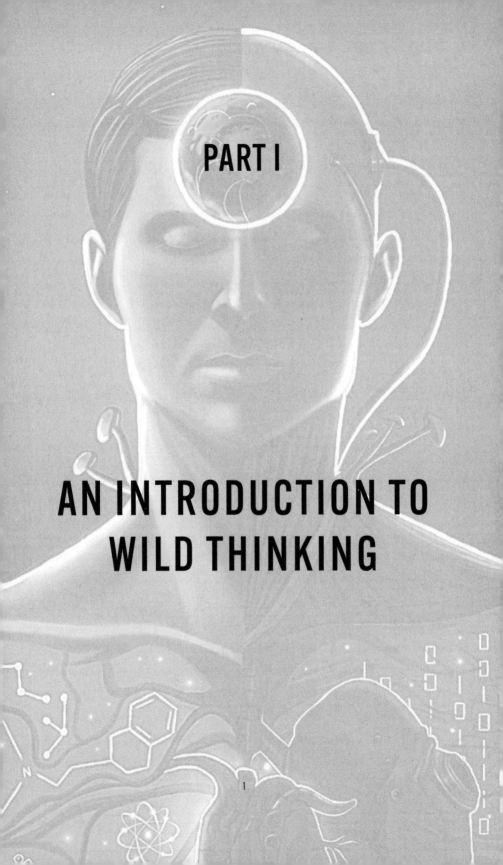

PART I

AN INTRODUCTION TO WILD THINKING

Chapter One

MOONSHOTS AS A FRAMEWORK FOR THE FUTURE

So, what exactly are moonshot moments?

Think of them as audacious leaps of faith, ambitious ideas that seem almost ludicrous at first glance. But if we can achieve such moments, we would completely reshape our world, challenging what humans think is possible. These aren't just incremental improvements; they're paradigm shifts that rewrite the script on what we thought humanity can achieve.

I believe moonshot moments are incredibly crucial now. The challenges our species faces today require a level of intentional cooperation and collaboration the likes of which we've rarely seen. Climate change and resource scarcity loom large. A growing population demanding more from a finite Earth. And technology, while offering incredible potential, also disrupts at an alarming rate. Business as usual just won't do. We need a new collective vision for the future and for our species, one that is neither dystopian nor rooted in outdated ideologies, but rather one that celebrates hyper-collaboration and innovates toward generating moonshot moments.

These moments appear when humanity is aligned in purpose and individuals are thriving. They signal that a different kind of thinking has challenged the status quo. They force us to stretch our minds, break free from the shackles of what's always been done, and envision a radically different future. Moonshot moments can be social, political, or even artistic endeavors that challenge the status quo and propel us forward.

As the founder of a venture studio, I spend my days guiding founders building companies at the forefront of technological innovation. But my focus isn't solely on building the latest applications for businesses or engineering new tech just for the sake of newness and profit. I'm driven by the potential of technology to improve the quality of life for humans and the longevity of our planet. This translates into building companies that harness the power of AI for social good in industries needing reform like education, incarceration, journalism, and the future of work. Some of the companies we're building at Ankord Labs, such as Narratives, Nota, and The Doe, are uplifting marginalized communities, creating empathy on the Internet, and sparking discourse on topics we otherwise avoid.

But fostering human progress requires more than just revolutionary ideas. History's most fertile periods of innovation, from the Renaissance to the Silicon Valley boom, weren't simply spontaneous eruptions of genius. They were fueled by a potent combination of factors, including investment and mentorship. The Renaissance-era de' Medicis, for instance, were not just patrons of the arts; they were shrewd investors who recognized the potential of groundbreaking artistic expression to elevate society.

Investors, tinkerers, and founders play a crucial role in creating a similar ecosystem for a potential future golden age. We act as incubators for pragmatic founders and creators, providing not just funding but also guidance, expertise, and a network of connections. This allows entrepreneurs to focus on their bold visions without getting bogged down by the administrative and logistical hurdles of scaling a company. In a way, we see ourselves as the modern-day de' Medicis, fostering a new renaissance of innovative thinking, artistic expression, creativity, and entrepreneurism. By nurturing these moonshot moments, we invest not just in groundbreaking companies but in a brighter future for all.

My passion for developing AI that is morally and socially responsible informs every step I take as an entrepreneur. That mission is expanded upon in this book. It's my way of igniting a vital discussion about the responsible development of coming technologies while also showcasing the excitement and promise of AI and other technologies under development. By learning from history's greatest successes, we can ensure that AI becomes a force for good, empowering individuals and driving positive change. The truth is, humanity's future isn't something that just happens to us—it's something we create. And with a little moonshot thinking, we can create something truly extraordinary.

Wild Thinking Is Not New

The call for moonshot moments isn't a desperate plea born out of our current anxieties; it's a call woven into the very fabric of human history. If you follow back through the ashes of our evolutionary history, you will see the trail lit by the fluttering light of rapid technological advancements. Living with fire was a shift brought about by our ancestors' bold curiosity and desire to venture beyond their comfort zone. They weren't content to just survive in the savanna. Imagine the very first glimmer of flame, a primordial dance of possibility and terror that would permanently change the path of human history. This leap of faith wasn't just about warmth or cooking; it unlocked new possibilities, from toolmaking to storytelling.

Fast-forward through the millennia, and you'll find humanity dotted with constellations of radical change. The agricultural revolution, a shift from nomadic hunter-gatherers to settled farmers, was no small feat. It required a radical reimagining of our relationship with the land. This gamble on the future forever changed the way we lived and interacted with our environment. Similarly, the taming of wild animals, from horses to wolves, wasn't just about companionship or beasts of burden; it reflected a daring spirit, a willingness to collaborate with creatures once considered adversaries.

History is a record of the advances in human technology, not just a dry compilation of dates and conflicts. Approximately between 800 and 200 BCE, the Axial Age witnessed a moonshot of the mind as philosophical and theological concepts blossomed across civilizations, challenging the nature of reality and our place in the world. The European Renaissance, an era of cultural renewal that lasted from 1450 to 1650 CE, was marked by an intense embracing of art, science, and human potential as well as a rejection of the status quo. The Age of Reason (1685 to 1815 CE), with its emphasis on logic and scientific inquiry, was yet another moonshot cluster, a shift in how we understood the world around us. We'll talk about these golden ages more soon.

These historical epochs weren't mere coincidence; they were periods when societies fostered the conditions for radical thinking to flourish. There was a convergence of elements—a spirit of cross-disciplinary cooperation, a tolerance for taking risks, and intellectual curiosity. We must continue to nurture these same components as we approach the dawn of a new era that calls for a unique constellation of moonshot moments.

As we examine the particular moonshot moments that have impacted our planet, keep in mind that they are living examples of the transformative potential of audacious ideation and interspecies cooperation, not artifacts from the past. They're a reminder that humanity has always thrived when we dared to dream big, work collaboratively, and aim to achieve the seemingly impossible. And in a world teetering on the edge of uncertainty, this spirit of free thinking isn't a luxury; it's a necessity.

Making Moonshot Clusters

It's impossible to deny the attraction of a startling, game-changing discovery —a bold idea or technology advancement that resolves an apparently intractable issue. But what happens if you can't wait for a fortunate break? The challenges we face, including climate change, resource scarcity, and technological disruption, demand proactive solutions. We can't simply stumble upon radical solutions when the need is so urgent. Instead, we need to cultivate the conditions that make these transformative breakthroughs more likely. It's kind of like laying down fertilizer to create a fertile ground where wild ideas can flourish in clusters, not as isolated occurrences.

Throughout history, we see confluences of moonshot moments—not just golden ages but periods of history I call "moonshot clusters"—that have been great for human progress. The moonshot cluster phenomenon can be attributed to a combination of factors. First, significant funding allows for the mobilization of large teams, cutting-edge technologies, and the exploration of diverse avenues simultaneously. Second, breakthroughs in technology can unlock the potential for multiple moonshot endeavors, creating a domino effect of innovation across various sectors. Lastly, when brilliant people work together, the combined knowledge of their many perspectives accelerates innovation and problem-solving, sometimes even challenging the status quo. This is why some of the biggest breakthroughs in history have come from coffee-drinking salons and conferences where philosophers, scientists, and inventors came to drink coffee and change the world. This unique combination of factors can lead to moonshot clusters, fundamentally altering existing systems and leaving a lasting impact on the world around us.

Encouraging full-fledged moonshot clusters means fostering a culture of hyper-cooperation, where individuals and organizations break down silos and work together. There is incredible potential for solving complex problems like climate change if scientists, engineers, and policymakers from around the globe could easily share resources and expertise.

But collaboration alone isn't enough. We also need to empower individuals to be active participants in this cognitive leap. This requires empowering individuals to capture and scale their own creativity and ingenuity by augmenting themselves with a personal knowledge management system that can be used to train their own AI agents. To broaden our mental horizons, we also need to take into account developments in transhumanism and psychedelic therapy. If everyone could nurture their own "spark" and add to the communal pool of innovative ideas, just think of the possibilities for creativity.

The bottom line? Inspirational moments are the outcome of conscious decisions and deliberations rather than a random lottery. Instead of waiting for a stroke of good fortune, we must deliberately foster the environments necessary for moonshot clusters and golden ages to flourish. By fostering collaboration, individual growth, and a spirit of bold thinking, we can transform the future, not simply hope for it.

But developing unconventional thinking and creativity involves more than just teamwork and self-determination. We'll have to negotiate and codify complex moral standards and intellectual frameworks as we develop new technologies and subsequently reinterpret what it means to be human. There are many potential pitfalls in transhumanism if we don't work hard to overcome our biases. We'll need to work toward promoting equity and protecting human rights. In a similar vein, space travel carries serious risks in the absence of careful environmental controls.

Our Moonshot Road Map

Before going further, it's worthwhile to examine the overarching structure of this book. This work is designed as a four-part exploration, with each section serving as a collection of central themes and ideas.

- **Part One:** An Introduction to Wild Thinking
- **Part Two:** Choice and Cooperation in the Face of Inevitable Change
- **Part Three:** Preparing the Self for Greater Creativity
- **Part Four:** Humanity's Next Frontiers: Embracing Our Cosmic Destiny

To ensure a cohesive journey through this intellectual landscape, I'll tie together the themes of each section in a short chapter-by-chapter synopsis before moving forward. These section summaries can help you locate the themes and arguments most important to you.

PART ONE:

An Introduction to Wild Thinking

We choose to go to the moon in this decade and do the other things, not because they are easy, but because they are hard, because that goal will serve to organize and measure the best of our energies and skills, because that challenge is one that we are willing to accept, one we are unwilling to postpone, and one which we intend to win, and the others, too.
—President John F. Kennedy, 1962

Part One is dedicated to unveiling the transformative power of moonshot moments. I'll show how these daring feats of creative genius, driven by a need to push limits and change the way we see the world, are fundamental to the story of humanity. The journey started with the introductory chapter you're reading right now.

We will examine the ethical implications and historical importance of transformative thinking in more detail in Chapter Two as we continue to explore the complexities of moonshot moments. We'll embark on a captivating voyage through five moonshot clusters marked by remarkable feats of human brilliance and groundbreaking change. Through this historical lens, we witness the profound impact of golden ages in shaping the course of our species and begin to consider how to best bring about our next golden age.

We'll also see how five key characteristics emerge as hallmarks of these transformative periods:

- Robust financial backing for new ideas
- A culture that embraces risk and experimentation
- A collaborative spirit that fosters synergy between diverse minds
- A celebration of unconventional perspectives
- A willingness to challenge established ways of thinking

These elements, acting in concert, create the fertile ground from which golden ages can spring.

PART TWO:

Choice and Cooperation

We are playing for the highest stakes in history. More souls are crammed onto this planet than ever, and there is the real prospect of commensurately great peril. At the same time, there is the prospect of building the infrastructure for a planetary first: enduring global concord. . . . And if we did that, if we laid a foundation for peace and fulfillment around the world, that would counterbalance a lot of past evils, given the number of people now around to enjoy the benefits. It may literally be within the power of our species to swing nature's moral scales—which for so long tended to equilibrate near dead even, at best—decisively in the direction of good.

—Robert Wright[1]

The intricate relationship between technical determinism and human agency is explored in detail in Part Two. Here, I confront the anxieties surrounding the impending arrival of artificial intelligence, synthetic biology, robotics, and other disruptive technological advancements.

You'll be thrown into the position of the architect of the future with a play in Chapter Three. With each scenario I'll ask you to picture yourself at a fork in the road, where each way points to a different direction for humanity. This choose-your-own-adventure style section will be your playground as you consider the consequences of your choices related to pandemics, cyber threats, and even alien encounters.

Certain routes will take you through utopian settings, thriving communities supported by bold thinking and groundbreaking innovations, but not every decision you make will result in sunshine and rainbows. Prepare yourself for more dire scenarios: worlds destroyed by widespread starvations or suffocating under the tyranny of autocratic governments. These terrifying scenarios serve as sobering warnings of what lies ahead if we choose not to adopt audacious concepts and cooperative measures. More than just a game, this immersive experience aims to make you painfully aware of the enormous influence that our collective decisions have on the world we inherit.

In Chapter Four, we'll return from a world of narrative play to engage with Mustafa Suleyman's techno-pessimistic viewpoint, which paints a picture of inevitable decline fueled by technological progress. Suleyman warns us about what he calls a metaphoric "coming wave" of tech that we cannot contain—a powerful convergence of AI, robotics, synthetic biology, and other technologies rapidly reshaping our world. As you'll see, many of my arguments throughout this section serve as powerful counterpoints to Suleyman, demonstrating that we possess the agency to navigate these advancements responsibly and avoid the most dire consequences.

I'll conclude the chapter by showing that core technologies such as artificial intelligence (AI) and synthetic biology are already a reality and are expanding quickly. Suppression attempts are futile and counterproductive, potentially pushing mavericks into unregulated spaces. I'll continually emphasize the need for proactive solutions—ethical considerations, responsible regulation, and continuous adaptation—to harness the potential of these technologies and navigate the inevitable wave of change they bring.

Our transition to an ethical technology development manifesto, which highlights the role that human choice plays in determining our future, will be furthered in Chapter Five. I'll debunk the notion of the lone genius and then promote teamwork as the catalyst for groundbreaking discoveries. I'll show that the secret to accomplishing bold objectives is hyper-cooperation, a win-win mindset promoted by author Robert Wright that encourages synergy between diverse minds. Through the adoption of hyper-cooperation and non-zero-sum thinking, we may maximize the advantages that will inevitably arise from the impending changes.

Chapter Five explores the biological underpinnings of hyper-cooperation in order to comprehend its efficacy. Examples of these include the symbiotic link between human cells and bacteria, the communicative capacity of fungus, the intricate social systems of insects, and the alluring lure of nectar. We'll discover that collaboration is ingrained in life itself. But biological predisposition is only the beginning. We must foster hyper-cooperation globally if we are to realize its full potential. Important steps in this quest include breaking down current power structures, overcoming cultural obstacles, and fostering trust between people and nations.

The difficulties are acknowledged throughout the chapter. Achieving genuine worldwide collaboration is an ambitious objective, considering current disputes, rivalries, and nationalisms. However, there are enormous potential benefits. Even technologies that appear to be individualistic, such as fire control, cooking, and mythology, are shown to be the results of extremely collaborative efforts.

PART THREE:
Preparing the Self for Greater Creativity

The "myself" which I am beginning to recognize, which I had forgotten but actually know better than anything else, goes far back beyond my childhood, beyond the time when adults confused me and tried to tell me that I was someone else; when, because they were bigger and stronger, they could terrify me with their imaginary fears and bewilder and outface me in the complicated game that I had not yet learned.

—Alan Watts[2]

Part Three embarks on a transformative journey dedicated to expanding the boundaries of both our brains and minds. This growth depends on the application of open, associational thinking, which is the process of combining seemingly unrelated concepts to produce a creatively insightful tapestry. Here, we shed the limitations of a strictly linear thought process and embrace a multidimensional landscape of intellectual exploration.

The exploration begins with a crucial step: acknowledging the inherent limitations of the human mind. Our egos often whisper an illusion of limitless knowledge, a delusion we must gently relinquish. In an age of information over-load, Chapter Six champions Personal Knowledge Management (PKM) as a tool for not just surviving but thriving. PKM empowers individuals to transform the deluge of information—articles, notes, social media—from a burden into a springboard for innovation.

The difficulty of information overload is acknowledged at the beginning of the chapter. With everything going on around us, it's difficult for us to concentrate and seize the fleeting moments of inspiration. In this case, PKM shows up as a lively knowledge haven where concepts can grow. By actively collecting, organizing, and reflecting on information, PKM fosters a deeper understanding and helps us identify connections—the sparks that ignite creativity, the seeds of moonshot moments.

I walk through some of the historical uses of meticulously crafted "commonplace books" to achieve a similar goal. These books served as a trusted companion, allowing the individual to readily access their accumulated knowledge for inspiration. The digital age has transformed these commonplace books into sophisticated PKM tools, offering powerful capabilities for organization and retrieval, but the precedent demonstrates the power of note-taking to foster associational thinking. The chapter then compares a variety of PKM approaches to help you select an option that is the best match for your personal mindset.

The topic of psychedelics' potential to transform mental health care and inspire moonshot ideas is covered in detail in Chapter Seven. I'll contest conventional wisdom on psychedelics, promoting scientific investigation into their potential therapeutic benefits for treating anxiety, depression, addiction, post-traumatic stress injuries, and other mental health conditions. I'll go into discussions with leading therapists and psychiatrists that will explain how psychedelics could revolutionize mental health care by providing alleviation when other approaches fall short.

The subjective experiences of psychedelic excursions, such as feelings of connection, ego dissolution, and access to long-forgotten memories, are also covered in this chapter. I take inspiration from the autobiographies of several "psychonauts," such as writer Michael Pollan and philosopher Alan Watts. In order to demonstrate the significant benefits of psychedelics for treating mental discomfort, my personal experiences are also intertwined.

The chapter ultimately concludes by acknowledging the controversy surrounding psychedelics but emphasizes the growing body of research that demonstrates their therapeutic potential and ability to foster creative thinking for solving global challenges. By disconnecting from the self, we open ourselves to a wellspring of previously unexplored experiences—spiritual, mental, and physical. This dissolution fosters a fertile ground for novel connections to form, fostering a more expansive view of reality.

Chapter Eight dives into the world of transhumanism, a movement that proposes leveraging technology to propel humanity beyond its current biological limitations. This school of thought views transhumanism as the next stage of human evolution, leveraging advances in implanted technology, biotechnology, and artificial intelligence to maximize human potential on the physical, cognitive, and emotional levels. Transhumanism's central tenet is that we are capable of surpassing our present constraints, including illness, age, and even death. But this vision extends far beyond individual augmentation. According to transhumanist ideals, these developments will benefit everyone in the future, resulting in a society that is more just and powerful.

By challenging our assumptions about what's possible, transhumanism inspires ambitious goals like radical life extension, biohacking, and superior intelligence. These objectives fuel research and development spending, greatly quickening the advancement of science. I'll give examples of biohackers like myself and show you cutting-edge transhumanist technology development trends and creative life-extending technologies from real-world situations throughout the chapter.

The chapter does, however, agree that the excitement surrounding transhumanism needs to be counterbalanced by serious thought given to its ethical ramifications and a removal of obstacles to accessing these technologies, which have too long been the preserve of the wealthy. We have to make sure that everyone who wants to benefit from these developments can, not just a wealthy few.

PART FOUR:
Humanity's Next Frontiers:
Transhumanism and Cosmic Destiny

Mother Nature, truly we are grateful for what you have made us. No doubt you did the best you could. However, with all due respect, we must say that you have in many ways done a poor job with the human constitution. You have made us vulnerable to disease and damage. You compel us to age and die—just as we're beginning to attain wisdom. You were miserly in the extent to which you gave us awareness of our somatic, cognitive, and emotional processes. You held out on us by giving the sharpest senses to other animals. You made us functional only under narrow environmental conditions. You gave us limited memory, poor impulse control, and tribalistic, xenophobic urges. And, you forgot to give us the operating manual for ourselves!

What you have made us is glorious, yet deeply flawed. You seem to have lost interest in our further evolution some 100,000 years ago. Or perhaps you have been biding your time, waiting for us to take the next step ourselves. Either way, we have reached our childhood's end.

We have decided that it is time to amend the human constitution.

We do not do this lightly, carelessly, or disrespectfully, but cautiously, intelligently, and in pursuit of excellence. We intend to make you proud of us. Over the coming decades we will pursue a series of changes to our own constitution, initiated with the tools of biotechnology guided by critical and creative thinking.

—Max More[3]

Part Four stands as a pivotal point in our exploration of innovative thinking, a daring venture into the realm of humanity's full cosmic potential. Here, we confront our responsibilities in regard to the enigmatic Fermi paradox, a gnawing silence that suggests we might be alone in this vast universe. This profound notion thrusts upon us a weighty mantle—the responsibility of not just understanding the universe but of becoming its shepherd. To fulfill this potential destiny, we must embark on a journey of self-discovery, pushing the boundaries of what it means to be human.

Chapter Nine focuses on humanity's grand leap toward space exploration, the ultimate moonshot endeavor. I explore the reasons why space exploration is no longer a luxury but a necessity driven by population growth and resource depletion on Earth. I draw heavily from futurists and theoreticians like Max Tegmark and Ray Kurzweil to inform my discussion of the physics behind expansion while tackling the moral questions of changing ourselves and the universe itself as we expand into the cosmos.

Potential answers to Earth's problems may lie in the immensity of space. I'll shed light on creative startups that are building the groundwork for mining asteroids in our solar system that may contain vital minerals, accessing water ice from moons like Europa, and using solar electricity in space with unparalleled efficiency. The prospect of extracting resources from space could give us hope for a time when humanity lives among the stars.

However, this chapter also examines how the human body loses its dominance in the hostile environment of space and how developing technologies can help humans cope with isolation, radiation, and microgravity. Artificial intelligence, for instance, can support scientific research, navigation, and decision-making in space missions. It is possible to design creatures using synthetic biology to produce fuel, food, and even building materials on distant worlds. A vision of human enhancement that blurs the boundaries between human and machine and enables us to endure the hardships of space flight is presented by transhumanism. I'll dissect these technological developments to better prepare us for space travel. And I'll acknowledge the ethical considerations of space exploration, emphasizing the importance of developing strong technomoral virtues to guide responsible development and use of these technologies.

The epilogue, Chapter Ten, serves as a pragmatic culmination of my exploration into humanity's boundless potential. We stand at a crossroads, facing not only the challenges of our time but also the thrilling possibilities that lie ahead. The Fermi paradox casts a long shadow, suggesting that if intelligent life is out there, it should be more readily apparent. Perhaps, then, we are the universe's best shot at galactic expansion, a responsibility that demands a surge in innovation and collaboration.

This final chapter brims with practical recommendations for amplifying our innovative spirit. For all of us, this journey we've begun toward moonshot moments isn't concluding. It's only beginning.

It's time for us to begin our long journey into human potential!

By understanding the importance of choice, fostering hyper-cooperation, upgrading our minds, and embracing a wider way of thinking, we can cultivate a culture that celebrates and nurtures transformative leaps. So, armed with these essential ingredients for moonshot thinking, let's begin our journey. We might just surprise ourselves with what we can achieve.

Chapter Two

HUMANITY'S HISTORY OF AUDACIOUS THINKING

Most kids dream of being an astronaut, but I was more than happy to keep my feet on the ground and my head in the clouds. Growing up with a Google executive for a parent in Silicon Valley during the early 2000s meant the Space Race felt like ancient history in comparison to the rapidly evolving realm of cyberspace. The transformation taking place in my pocket was much more fascinating than humanity's unquestionably spectacular journey to the skies. Not the huge void of space, but the uncharted territory of the little instrument that was attached to my palm was the most important frontier in my eyes.

My dad, a constant harbinger of the future, would bring home prototype phones from work. I remember clearly the day he upgraded from his cumbersome Blackberry to the G1 (commonly known as the HTC Dream). For me, this miracle was a moonshot moment, quickly followed by the Nexus 1, the first stylish touchscreen Android phone without a keyboard. Before long, my father was teasingly annoyed when my sister and I, captivated by the iPhone's svelte form, easy-to-use interface, and popularity among young people, embraced the device wholeheartedly. The future appeared to be curled up in the palm of my hand rather than existing outside it.

Unlike the generation that worshipped Neil Armstrong and sang along with Elton John's "Rocket Man," my teenage heroes tucked their T-shirts into their jeans and walked across stages in Silicon Valley to give keynote presentations. I'd race home from school, heart pounding with anticipation, to witness the latest unveiling by Steve Jobs. I'd watch the videos again and again, and then later I'd immerse myself in the review videos on YouTube. The thrill of witnessing a new technological frontier unfold became a year-round obsession. Google or Apple launch days were marked on my calendar as religious holidays. I consumed every YouTube review video with a fervent devotion, from exhilarating anticipation to post-release analysis.

The small advancements made from generation to generation were evidence of the never-ending quest for innovation, not just insignificant minutiae. Not only was phone technology a toy, but it was also my first real moonshot experience, a life-changing adventure that I saw coming together year after year, update after update.

Steve Jobs felt, to me, like a modern-day Gutenberg. The development of the cell phone may appear to be very different from the revolutionary effects of the printing press at first glance. Ultimately, one enabled instantaneous communication while the other ushered in a mass-production era of information. But upon deeper examination, it became clear that both innovations spurred a wave of social and cultural shifts in addition to streamlining already-existing chores.

Like the printing press, which democratized access to knowledge and forever altered the landscape of education and scholarship, the cell phone's evolution from an awkward brick to a ubiquitous supercomputer has profoundly changed how we engage with the world. It's about having a continual connection, being able to get information instantly, and being able to communicate with people instantly—it's not only about replacing landlines and pagers. Our daily lives, the way we form connections, how we capture our thoughts, and how we consume news have all been profoundly changed by this hyper-connectivity.

Gutenberg's printing press may have laid the groundwork for the information age, with all its attendant joys and complexities, but the cell phone has truly propelled us into it. And when I think of moonshot moments, the evolution of the cell phone from telephone to an AI-wielding communications powerhouse is the giant leap I witnessed and embraced. As a moonshot moment, the explosive development of the cell phone ignited a passion for technology that continues to shape my life today. I've even converted my iterative cell phone collection into a modern art wall decoration, so that I can remember the value of wild thinking and iterative improvement!

Taking a Gamble: The Origins of Moonshot Moments

But the idea of moonshot moments didn't start with Jobs or the cell phone. It was born from humanity's cosmic yearning. Inspired by the utterly outrageous goal of landing humans on the moon, the term *moonshot* was adopted to describe any exceptionally ambitious and challenging project. This association with the Space Age imbued the term with a sense of innovation, daring, and the pursuit of seemingly impossible goals. It captured the spirit of pushing boundaries and venturing beyond the realm of the ordinary. Despite the inherent risks and uncertainties, the Apollo missions (1961–1975) embodied the core spirit of a moonshot: unwavering ambition, a willingness to embrace the unknown, and the pursuit of a seemingly impossible dream.[1] The program's success, which was attained through years of unwavering work, teamwork, and creativity, not only represented an important turning point in human history but also solidified the term *moonshot* as a potent representation of daring undertakings and the possibility of groundbreaking accomplishments.

In recent decades, the term *moonshot* has spread beyond its spacefaring origins to be employed in a wide range of contexts. These days, it's commonly used in the corporate world to describe creative and disruptive initiatives that aim to solve complex problems or provide groundbreaking results.[2] This expression stresses setting lofty goals and having a significant impact in a range of contexts, such as environmental sustainability, social change, and healthcare.

Though the term's meaning has evolved over time, its core concepts of creativity, ambition, risk-taking, and the determination to challenge the status quo still remain. It serves as a powerful reminder that anything is possible, even seemingly unachievable objectives, if one has vision, creativity, and unwavering determination.

Moonshot moments are those revolutionary discoveries, concepts, fads, or happenings that daringly alter the path of human history. Often pushing the edges of what is currently thought to be attainable, they are characterized by a distinct blend of ambition, unusual thinking, and revolutionary potential. This drive necessitates unconventional approaches. Mavericks and renegades prosper. The ultimate aim of a moonshot goes beyond succeeding; it seeks to transform the landscape by creating significant, lasting impact that revolutionizes entire fields or addresses critical global challenges.

Moonshot moments are all about momentum. Like a blizzard on a mountainside, they come suddenly, then come harder and harder until they collect momentum and soon you've got an avalanche. That's the impact of moonshot clusters. This cascade effect of theirs stands in stark contrast to incremental improvements. While incremental improvements involve gradual, iterative advancements within existing frameworks, moonshot clusters often come in rapid bursts of innovation.

By examining specific historical eras of moonshot clusters, we can gain a deeper understanding of the factors that contribute to bursts of innovation that have shaped and altered the course of human history. With hope, we can also determine what factors would help us create more moonshot clusters in the future.

The Philosophical Underpinnings of Moonshot Moments

A number of philosophical frameworks shed light on why moonshot moments are so significant to us as a species. As a practicing Stoic myself, that's the perspective I start with. Moonshot projects stand for the quest of greatness and morality in our eyes. By setting high standards for ourselves and overcoming obstacles, we develop fortitude, bravery, and discernment. These endeavors showcase the power of human determination and inspire us to strive for excellence in all aspects of our lives.

But with its focus on maximizing overall happiness and well-being, utilitarianism offers a better framework for evaluating the merits of moonshot moments. Proponents of this philosophy might argue that moonshot endeavors are justified if they have the potential to significantly improve the lives of many people. This could involve breakthroughs in healthcare, advancements in renewable energy, or innovations that address global challenges like climate change, poverty, or hunger. But utilitarians would also urge careful consideration of the potential downsides and unintended consequences of wild thinking. The resources and risks involved in such ambitious projects need to be weighed against the potential benefits to ensure they align with the principle of maximizing overall well-being. This utilitarian lens encourages a balanced approach to moonshot moments, urging both bold vision and responsible evaluation of their impact on society.

Moonshot moments are sometimes romanticized as bursts of pure invention, conjured from the ether by visionary geniuses. Eureka! But the reality is far messier and often grounded in the practical philosophy of pragmatism. This approach places more emphasis on the value of experimental and practical work than it does on theoretical thinking. Pragmatists ask "what works" rather than getting bogged down in the "what ifs." Iterative processes enable moonshot ideas to be evaluated, improved, and eventually realized.

After all, Thomas Edison once famously remarked, "I have not failed 10,000 times—I have successfully found 10,000 ways that will not work." It's the difference between dreaming of flying cars and building functional prototypes, no matter how crude they may be initially. Pragmatism offers a rich environment for growing the seeds of ambitious ideas into workable solutions by accepting the messy realities of trial and error.

At the heart of many moonshot moments lies a powerful approach called first principles thinking. Originating in pre-Stoic ancient Greece and made popular by Aristotle, first principles thinking isn't about accepting the status quo or simply iterating on existing ideas. Instead, it's about stripping a problem down to its bare essentials, questioning fundamental assumptions, and rebuilding solutions from the ground up. It's akin to a scientist questioning long-held beliefs about the natural world or an artist reimagining the very essence of form and expression. This radical approach is what allows moonshot moments to emerge. By refusing to be bound by limitations or entrenched practices, first principles thinking creates the fertile ground in which audacious ideas and disruptive innovations can take root. It's the spark that ignites the moonshot—the willingness to challenge the very foundation of what's deemed possible and forge a new path toward a transformative future.

A different, existentialist viewpoint is provided by Søren Kierkegaard, who views the moonshot through the prism of experimentation and iteration. For Kierkegaard, the moonshot would represent a leap of faith,[3] a decisive commitment to a seemingly irrational pursuit in defiance of established norms and limitations. This leap isn't driven by cold logic or guaranteed success but by a passionate conviction in the face of uncertainty. It's the astronaut choosing to embark on a perilous journey into the unknown, driven by a yearning to push the boundaries of human experience. This Kierkegaardian lens emphasizes the courage and personal agency inherent in moonshot endeavors, highlighting the bold spirit that fuels humanity's drive to reach beyond the seemingly impossible.

Moonshot moments, with their bold objectives and capacity to change the world, seem to me to be very closely aligned with the fundamental principles of existentialism. Existentialists like Jean-Paul Sartre and Kierkegaard held that "existence precedes essence," meaning that, rather than being born with a predetermined purpose, people construct their own meaning by their choices and actions in a chaotic world. Moonshot endeavors, by definition, force people to deliberately create meaning by encouraging them to aim for the seemingly unachievable and to face the fundamental absurdity of the cosmos. To combat existential threats like climate change, I think we have to choose wild thinking in a resource-limited world. This endeavor, full with uncertainty and the possibility of failure, embodies the existentialist idea of personal accountability and the bravery to define one's own existence.

The emphasis on the will to power in Friedrich Nietzsche's philosophy provides a distinctive viewpoint on moonshot moments. According to Nietzsche, people and society are always trying to get past obstacles and establish their authority. In this context, moonshot initiatives become manifestations of this underlying motivation. They stand for power directed at accomplishing the seemingly unachievable, stretching the bounds of what is seen as possible, and completely altering the concept of what is conceivable. Whatever the result, attempting a moonshot reflects the bold attitude of self-actualization that Nietzsche espoused. It's the entrepreneur taking on industry heavyweights with a disruptive breakthrough, the scientist upending the status quo with a novel hypothesis, or the artist exploring uncharted creative territory. Moonshot moments become monuments to the human ability to pursue mastery and redefine what is possible when viewed through the prism of Nietzsche's philosophy.

Through the application of these many philosophical vantage points, we are better able to understand the complex interplay between practical utility, social responsibility, and norm violation that underlies many moonshot moments. They leave a lasting mark on our collective imagination, embodying not only scientific wonders but also our hopes for meaning, self-overcoming, and beneficial societal impact. Although philosophical viewpoints provide useful frameworks for comprehending the goals and motives that underlie moonshot moments, a deeper comprehension depends on looking at the historical setting in which they occur.

Each moonshot moment is like a unique puzzle piece, fitting into the bigger story of humanity's journey: a blip on a long march forward. By exploring the specific circumstances that birthed these bold ideas, we can move beyond dry theory and connect with the real people, the real struggles, and the real triumphs that make moonshots so darn inspiring. So, let's ditch Nietzsche (he never minds) and get our hands dirty with the fascinating history behind these game-changers!

Humanity Breaks Out: Moonshot Moments in the Ancient World

This was the reality for much of human history: we were human for more than 100,000 years before we ever built a structure that's still standing.

Early humanity began in scattered communities living primarily as nomadic hunters and gatherers, with limited knowledge passed down through generations. However, around 12,000 years ago, a revolution began to unfold. The transition to sedentary life in permanent settlements marked a pivotal turning point. These nascent urban centers, like Çatalhöyük in Anatolia and Jericho in the Levant, fostered closer social interaction, knowledge sharing, and specialization. This fertile ground became the foundation for a surge of innovation that forever altered the course of human history.

The ancient world was, for centuries before 1000 BCE, focused in fertile river basins in Egypt, the Middle East, China, and India, where the elements controlled humanity and not the reverse. That began to change when humans took control of the rocks at our feet. Transforming readily available materials like copper and tin into sturdier tools and weapons marked a significant leap forward for humanity. Around 4000 BCE, early civilizations in both Mesopotamia and the Indus Valley pioneered metalworking techniques through trial and error, discovering that heating these materials and hammering them into specific shapes yielded stronger and more versatile tools. This breakthrough ushered in the Copper Age and later the Bronze Age, enabling the creation of sharper agricultural implements for increased food production, sturdy building materials for more permanent structures, and effective weapons for defense and conquest. The capacity to work with and shape metals stimulated artistic expression by enabling the creation of elaborate jewelry and sculptures in addition to revolutionizing utilitarian applications.

A moonshot moment in metals was well underway.

The wheel is another example of the revolutionary force of human inventiveness. It was independently created in Mesopotamia and India between 3500 and 4500 BCE, respectively. This seemingly straightforward invention—a circular disk turning on an axle—revolutionized almost every aspect of daily life. It made it possible to move people and goods across great distances with efficiency, which aided in the development of transcontinental commerce networks. The invention of the wheel influenced architecture, agriculture, and warfare by accelerating the development of complex vehicles like chariots and wagons. Its impact went beyond the material world as it came to represent innovation and development in many cultural contexts.

And if the wheel meant trade, trade meant writing. It's hard to imagine a world where knowledge could only be passed down through oral traditions, susceptible to distortion and loss over time. But that was a significant portion of human history. The ability to capture and transmit information beyond the spoken word represented a monumental achievement and was a necessity for long-distance trade. By 3500 BCE, the Sumerians in Mesopotamia had developed cuneiform, a system of wedge-shaped impressions on clay tablets, meticulously recording their history, laws, and economic transactions (along with their complaints about one another[4] and their bar jokes[5]). Egyptians devised hieroglyphics, a combination of pictographs and ideograms, adorning the walls of tombs and temples with intricate narratives and religious imagery. The Indus Valley civilization is believed to have developed its own undeciphered writing system around the same time. These writing systems not only facilitated record-keeping and communication across vast distances but also preserved cultural heritage and scientific knowledge for future generations.[6] In fact, the development of languages over centuries caused changes in our brain processing to allow us to handle symbology more efficiently.[7] The ability to document and share ideas across space and time proved to be a powerful catalyst for further innovation and societal advancement.

For early civilizations, taming the unpredictable character of water through reservoirs and canal networks proved revolutionary. Rainfall reliance reduced agricultural output prior to 4500 BCE, and communities were limited to places with reliable water supplies. In order to manage the yearly floods and disperse water across large areas of land, the Egyptians living along the Nile River constructed an intricate network of canals. The Mesopotamians in the Tigris and Euphrates river valleys employed similar techniques, constructing intricate networks of canals and reservoirs to irrigate their fertile crescent. The development of irrigation systems not only revolutionized agriculture but also laid the groundwork for urban planning and infrastructure development. These advancements enabled early civilizations to cultivate larger areas of land, ensuring food security for growing populations and fostering the rise of complex societies with specialized roles and social hierarchies.

Many of these moonshot moments seem so basic and obvious now, but like a child discovering how to build towers of blocks, we had to begin with sturdy foundations. Despite its limitations, the ancient world laid the groundwork for future advancements as we see the rise of cities and organized societies. The convergence of urban life, knowledge sharing, and a spirit of exploration fostered a unique environment conducive to moonshot moments. These innovations laid the foundation for future breakthroughs, shaping the trajectory of human civilization for millennia to come. In particular, these early moonshot moments offer humanity our first opportunities to take control of the elements and natural materials around us and use them in extraordinary new ways.

Other significant moonshot achievements made by the ancient world include harnessing fire, creating agriculture, and populating cities, but we'll get to those later. For the time being, the ancient world's emphasis on pragmatic problem-solving, its spirit of cooperation, and its openness to experimentation serve as reminders of the persistent human capacity for creativity, motivating us to keep pushing the envelope of what is currently feasible.

Spiritual Awakening: Intellectual Ferment and the Axial Age

The Axial Age, roughly from 800 to 200 BCE, was a critical juncture in human history characterized by a remarkable upsurge in philosophical and intellectual activity among various civilizations. This period saw a "new awakening of human consciousness,"[8] as philosopher Karl Jaspers put it, characterized by a move toward introspection, ethical reflection, and the search for ultimate reality. This golden age offered the ideal environment for a number of moonshot events that profoundly altered our perceptions of the spiritual cosmos and our place within it. These moments were driven by growing urbanization, trade, and cultural exchange.

Jaspers argues that during the Axial Age, humanity reached a new level of cognitive development, enabling individuals to think abstractly, contemplate existential questions, and challenge established norms. This expanded capacity for thought, along with the idea sharing that came from greater cross-cultural interaction, created the groundwork for revolutionary movements in philosophy and religion.[9] Other authors, like renowned religious historian Karen Armstrong, have embraced the notion of the Axial Age to demarcate a period of great religious transformation in human consciousness.[10]

There are numerous legitimate critiques of Axial Age theory, some even claiming that Jaspers has put so much overemphasis on intellectual advancements during the period that he's created a myth that the last millennium BCE was an exceptional time when it really wasn't.[11] But to me, while that last millennium is not a clearly defined technological period,[12] and there is not necessarily a tie between all the intellectual movements that sprung up around the world at this time,[13] the sheer scope and range of intellectual moonshot moments during the last millennium BCE cannot be ignored. Axial or not, the last millennium before the birth of Jesus was definitely a moonshot cluster of deep intellectual impact.

Emerging in ancient Persia around 1000 BCE, Zoroastrianism introduced a revolutionary concept that challenged the prevailing polytheistic beliefs of the time. It proposed the existence of a single, all-powerful God, Ahura Mazda, locked in an eternal struggle with the forces of evil embodied by Angra Mainyu. This novel idea not only had profound theological implications but also laid the groundwork for the development of ethical and moral frameworks that emphasized individual responsibility, righteous living, and social justice.[14]

Zoroastrianism's influence extended beyond its immediate geographical reach,[15] impacting the development of later Abrahamic religions like Judaism, which emerged in the Levant around 600 BCE. Judaism further developed the concept of monotheism, establishing Yahweh as the sole deity and emphasizing a covenant between God and the chosen people.[16] These moonshot moments in religious thought not only reshaped spiritual practices and societal structures but also laid the foundation for ethical and philosophical discourse that continues to influence the world today.

Two separate philosophical schools that would have a significant impact on Chinese history and culture originated in China around the sixth century BCE. Confucius promoted a community ruled by decency, propriety, and ceremony, stressing the value of social harmony, filial piety, and moral behavior. His teachings, which are collected in the Analects, placed a strong emphasis on the creation of a just and peaceful society through education, self-cultivation, and sound administration. The creator of Daoism, Laozi, put out an alternative philosophy based on adhering to the Dao's natural order, or the "way." Daoism promoted connecting people with the natural flow of the universe by emphasizing simplicity, spontaneity, and inaction. These moonshot moments in Chinese thought not only provided frameworks for individual conduct and societal organization but also permeated diverse aspects of Chinese culture, from art and literature to politics and social practices.

We can also travel to Axial Age India in the sixth century BCE, when Siddhartha Gautama, often known as the Buddha, presented the Four Noble Truths and the Eightfold Path as a revolutionary means of achieving emancipation from suffering. The Buddha's new religious movement opposed the dominant caste structure and highlighted how meditation, self-reflection, and moral behavior could lead to enlightenment for everyone. As Buddhist teachings proliferated throughout Asia, they had an impact on societal institutions, literature, art, and spiritual activities. The establishment of monasteries as centers of learning and scholarship further fostered the dissemination of Buddhist thought and cultural exchange across vast geographical regions.[17] This moonshot moment in religious thought continues to influence millions of followers worldwide, offering a unique perspective on human existence and the pursuit of ultimate liberation.

As an alternative to religious thought in the fifth and fourth centuries BCE, ancient Greece witnessed the flourishing of philosophy as a systematic approach to understanding the world. This intellectual revolution marked a significant departure from traditional myth-based explanations and ushered in an era of rational inquiry and critical thinking.[18] Thinkers like Socrates, with his dialectical questioning technique, questioned the status quo and made people rethink their presumptions and beliefs. Then, building on these foundations, Plato produced works such as *The Republic* and *The Symposium* that developed conceptions of reality, knowledge, and the ideal state. Aristotle, a dedicated student of Plato, increased the scope of philosophical investigation by stressing the use of reason, logic, and empirical observation in comprehending the natural world. These seminal moments in the history of ideas established the foundation for Western philosophical tradition, producing a long-lasting impact on political theory, scientific research, and ethical debate for centuries to come.

All of these intellectual movements got a great deal of traction partly because of the empire, a new moonshot strategy for spreading knowledge across ever greater numbers of people. Empires promoted extraordinary levels of trade, infrastructural development, and cultural interchange across their huge geographic expanses and heterogeneous populations.

For example, the Achaemenid Empire, established by Cyrus the Great in the sixth century BCE, implemented innovative administrative systems and standardized communication with Aramaic script, facilitating the exchange of goods, ideas, and cultural practices across its expansive domain. Cyrus also added the notion that empires would rule over diverse people and bring all of their ideas into a melting pot, and "imperial ideology from Cyrus onward has tended to be inclusive and all-encompassing."[19]

Similarly, the Mauryan Empire, founded by Chandragupta Maurya in India in the fourth century BCE, promoted the spread of Buddhism throughout South Asia and established a centralized administration that facilitated trade and infrastructure development.[20]

The Han Dynasty, which came to power in China in the third century BCE, brought with it an era of increased economic prosperity as well as advances in technology and artistic expression. The legendary Silk Road, which linked China and the West, was an essential means of trade and cultural interchange that promoted the spread of ideas, artistic movements, and religious convictions over great distances. These empires contributed to the Axial Age's richer fabric of human experience by consolidating political authority, fostering economic expansion, and acting as testing grounds for new ideas and cultural innovations.

Despite vast differences in language, culture, and prior belief systems, civilizations in Persia, Greece, India, and China all independently grappled with fundamental questions about existence, morality, and the nature of reality during the Axial Age. This parallel blossoming of intellectual inquiry underscores the universality of certain human concerns and the inherent drive to seek meaning and understanding. While the specific expressions of these ideas—from the monotheistic philosophies of Zoroastrianism and Judaism to the introspective path of Buddhism to the structured logic of Greek philosophy—varied greatly, each offered a unique lens through which to understand the human experience. This diversity of perspectives adds immense value to the tapestry of human thought, enriching our collective understanding of ourselves and the world around us. The Axial Age serves as a powerful reminder that shared human aspirations, coupled with the unique cultural contexts in which they flourish, can generate a spectrum of intellectual riches that broaden and deepen our understanding of the human condition.

Whether the period is a clearly delineated Axial Age is irrelevant to me: It's a moonshot cluster of the most spiritual, most philosophical kind. It's what we'll need to survive the challenges that lay ahead of us.

Lights in the Desert: The Persian and Arab Golden Ages

While Europe grappled with instability and fragmentation after the fall of the Roman Empire and the Early Middle Ages—the "Dark Ages"—other parts of the world witnessed remarkable periods of intellectual and cultural flourishing. The Islamic world experienced a pair of golden ages from the eighth to the thirteenth centuries CE, marked by groundbreaking contributions in various fields, including mathematics, astronomy, medicine, and engineering. These golden ages don't belong to a singular empire but rather a constellation of power centers that fostered a vibrant exchange of ideas.

The interactions between empires and regions were fascinating to observe throughout the Arab world. With Baghdad serving as its capital, the Abbasid Caliphate grew to become a formidable force. After being first brought together by their shared Islamic faith, the Abbasids zealously translated and distributed information from throughout antiquity while also adopting a more centralized administrative structure. Across the Mediterranean, Arabs flourished and established themselves in Morocco and Southern Spain. Al-Andalus, as this region is known, developed into a center of thought and culture. Scholars who were Muslim, Christian, and Jewish coexisted and worked together to enhance astronomy, medicine, and mathematics. This period of relative tolerance stood in stark contrast to the religious tensions brewing in Europe, where the Crusades, a series of holy wars launched by Christian powers to reclaim Jerusalem, exemplified the growing religious and political divisions of the time.

Simultaneous golden ages in both of these epicenters of the Arab world meant a period of genius-driven science, as well as a time built around the brilliant idea that it would be great to ensure the transmission and preservation of earlier ancient and Axial Age genius ideas! Established in Baghdad by the Abbasid Caliphate in the eighth century CE, the "House of Wisdom" served as a pivotal center for scholarship, translation, and the dissemination of knowledge. Scholars from diverse backgrounds, including Greeks, Persians, Indians, and Christians, gathered at the House of Wisdom to translate and synthesize vast amounts of scientific and philosophical texts from various civilizations. Think of the House of Wisdom as the world's first photocopier, at a time when a single fire could wipe out entire authors' contributions to discourse. We would not still have the works of Greek philosophy in particular if it were not for the House of

Wisdom's scroll copiers translating them to Arabic and Farsi. This translation movement played a crucial role in preserving and transmitting classical knowledge to the Islamic world and beyond, laying the foundation for future scientific and intellectual advancements.

During the Arab Golden Age, Islamic physicians made great advances in the fields of surgery and medicine, building on the foundations set by previous civilizations. They built hospitals as hubs for patient care, medical research, and education, such as the well-known Bimaristan in Baghdad. Furthermore, Arab doctors such as Al-Zahrawi (Albucasis) invented cauterization and sutures, among other surgical techniques, and recorded their methods in illustrated medical treatises.[21] Through translations and cross-cultural interactions, these developments not only enhanced healthcare procedures in the Islamic world but also had a significant impact on medical knowledge and practices in Europe. These developments marked a radical break from earlier medical interventions that depended as much on wishful thinking as any physical measurements.

Often referred to as Alhazen in the West, the Iraqi scholar Ibn al-Haytham made significant contributions to the field of optics during the tenth and eleventh centuries CE. His groundbreaking work, *The Book of Optics*, challenged prevailing theories of vision and laid the foundation for modern optics. Ibn al-Haytham conducted pioneering experiments on light and vision, introducing concepts like refraction, reflection, and the camera obscura. These advancements not only revolutionized our understanding of light and vision but also had lasting implications for the development of telescopes, microscopes, and other optical instruments.

Similarly, the Persian empires of this era, like the Samanids and the Buyids, were vital threads woven into the rich tapestry of the Persian Golden Age. These empires maintained a strong sense of their unique political and cultural traditions even as they joined the Arab world in embracing Islam. The Samanids became well-known as intellectuals' and scientists' patrons, and they were based at the strategically important crossroads of Central Asia. Their wise guidance encouraged groundbreaking studies in astronomy, medicine, and mathematics, advancing these disciplines. By contrast, a more decentralized government was developed by the Buyids, who ruled from Persia's center. However, their focus on trade and diplomacy proved equally crucial. By facilitating the exchange of goods and ideas across the region, they nurtured a vibrant intellectual and cultural exchange that further enriched the golden age. This interplay between centralized patronage and decentralized trade routes exemplifies the multifaceted nature of this remarkable period in history.

Persian scholars like Al-Razi (Rhazes) and Ibn Sina (Avicenna) made significant contributions to diagnosis, treatment, and pharmacology.[22] Unlike his predecessors who relied solely on established theories, Al-Razi championed the importance of clinical observation and meticulous record-keeping. This (for the time) wild approach, emphasizing firsthand experience over blind acceptance of tradition, fueled groundbreaking discoveries. His most notable contribution was the compilation of a vast medical encyclopedia, known as *The Comprehensive Book of Medicine*. It was a thorough analysis of diseases, incorporating al-Razi's own observations and clinical experiences. The encyclopedia served as a reference text for centuries, not only in the Islamic world but also in Europe, influencing generations of physicians with its innovative categorizations of illnesses, detailed descriptions of symptoms, and groundbreaking treatments based on al-Razi's meticulous research. His work on smallpox and measles,[23] for instance, stands as a testament to his dedication to clinical observation and remains a cornerstone of our understanding of these infectious diseases.

The polymath Ibn Sina, or Avicenna in the West, came from the dynamic intellectual environment of tenth century Persia and made a lasting impact on the history of medicine.

His comprehensive medical text, *The Canon of Medicine*, was an organized study of the human body that covered anatomy, physiology, and the management of a wide range of illnesses. Ibn Sina offered novel theories regarding the origins and manifestations of diseases, relying on the medicinal traditions of Greece, India, and Persia. Generations of doctors have relied on his groundbreaking views on pulse diagnosis and the application of psychological variables in recovery.[24] *The Canon of Medicine* transcended geographical and cultural boundaries, becoming a cornerstone of medical education in both the Islamic world and Europe for centuries to come.

The ninth century CE witnessed a pivotal moment in the history of mathematics with the groundbreaking work of Muhammad ibn Musa al-Khwarizmi, a brilliant mathematician and polymath from Persia. His seminal work, *The Compendium on Calculation by Completion and Balancing*, introduced the concept of algebra to the world. Algebra was a new mathematical concept and a fundamentally new approach to working with numbers. Al-Khwarizmi's work presented a systematic method for solving mathematical equations using variables and unknowns, a revolutionary approach that replaced the cumbersome geometrical methods previously employed.

The Persian and Arab golden ages stand as testaments to the transformative power of intellectual curiosity, celebration of diversity and collaboration, and a commitment to continuous learning. The wide bridge of the Islamic world, from southern Spain to Persia, was fostered by a deep respect for knowledge and a culture of open exchange. These societies valued the iterative character of science, expanding on what was already known and continuously aiming to enhance and improve upon earlier findings. This commitment to creativity and experimentation turned out to be crucial in producing a cluster of moonshot moments that not only revolutionized their own cultures but also had a cascading effect on human history worldwide. These moonshot moments serve as powerful reminders of the immense potential that can be unlocked when diverse people of mixed backgrounds come together in a spirit of collaboration and continuous learning.

Beauty over Brute Force: The European Renaissance

Sometimes, I think it's a bit unfair to call the European Renaissance (fourteenth through seventeenth centuries) a "renaissance" at all, because the European Renaissance wasn't some phoenix, some reboot of the Roman Empire. Rather, it was a completely transformative explosion of creativity and intellectual inquiry. Striking a new path after the devastation of the Black Death and the petering out of the Crusades, this period witnessed an extraordinary cluster of moonshot moments that fundamentally reshaped Western civilization, leaving the Middle Ages in the dust and laying the groundwork for the modern world. It was as if humanity, having stared into the abyss, collectively decided to invent the light bulb.

The Italian city-states, especially Florence, Venice, Milan, and Rome, became thriving creative hotspots thanks to the support of affluent families that desired not only power but also prestige. Despite their political maneuvering, the de' Medicis and Borgias considered themselves to be patrons of the arts, understanding that these pursuits could enhance their reputations and create a lasting legacy that extended beyond wealth and land. Rich enough to match their aspirations, these benefactors lavishly funded the creations of visionary artists such as Michelangelo, Leonardo da Vinci, Raphael, and Botticelli. Their paintings and sculptures, characterized by a rediscovery of classical forms and a focus on human anatomy and perspective, were explosions of creativity that revolutionized the visual arts and continue to inspire generations today. These artists, armed with brushes and chisels instead of swords and lances, conquered the hearts and minds of the world, proving that beauty could be just as powerful as brute force.

The political machinations and rivalries between these city-states also inadvertently fostered innovation. Prototypes of tanks and flying aircraft that Leonardo da Vinci drew were ordered by nervous clients looking for the most advanced weaponry, not for aesthetic purposes. The constant need for defensive fortifications and advancements in weaponry spurred engineering and architectural breakthroughs. This spirit of innovation extended beyond the battlefield, leading to moonshot moments in diverse fields, from city planning with the development of ideal cities like Sforzinda to religious reform with the Protestant Reformation. These transformative changes not only reshaped the European landscape but also set the stage for the Age of Exploration and the scientific revolution, propelling humanity toward a new era of discovery and understanding.

Before the Renaissance, paintings were more like maps than windows into another world. Artists primarily relied on flat, two-dimensional representations, leaving viewers to imagine the depth and dimension beyond the canvas. Go to the Medieval Art section of any major museum, and you'll see a lot of stern, flat-faced Marys looking very confused about their weird fingers!

Thankfully, the fourteenth through fifteenth centuries meant the idea of linear perspective, a game-changer pioneered by the likes of Filippo Brunelleschi and Alberti. I'm not just talking about drawing straight lines but rather about shattering the confines of the flat canvas and creating the illusion of depth and realism. Suddenly, paintings could transport viewers to bustling marketplaces, sprawling landscapes, or even the divine heavens, all thanks to the magic of converging lines and vanishing points. This moonshot moment in art forever altered the course of Western painting, allowing artists to depict scenes with a breathtaking sense of spatial organization and naturalism that continues to captivate audiences today.

The Renaissance elevated polymaths to a position of prominence and promoted interdisciplinary thinking before it was fashionable. Prodigies such as the quintessential Renaissance man, Leonardo da Vinci, defied classification, showing remarkable aptitude in the fields of anatomy, engineering, sculpting, and painting. His boundless inventiveness and curiosity left a legacy that still inspires wonder decades later. Another titan of the age, Michelangelo, transformed the possibilities for sculpting and fresco painting with his famous pieces, such as the Sistine Chapel ceiling and David. The sheer size and emotional impact of his work left spectators in awe.

Beyond the edges of the Italian peninsula, we'd be remiss to overlook William Shakespeare, the Bard of Avon. His classic plays, such as *Hamlet* and *Romeo and Juliet,* are still performed and studied today because their themes and characters continue to speak to audiences of all ages and cultures. Not only were these literary and creative titans producing works of art, but they were also questioning social mores, igniting the imagination of future generations, and demonstrating the limitless potential of humankind.

The Renaissance marked a significant change in the ways that philosophy, religion, politics, art, and literature viewed humanity. Its undergirding principle was humanism, a philosophical movement that ventured to give individuality—rather than just the divine—center stage. This was about realizing the potential and dignity that each and every human being possesses, and not about overthrowing God. This anthropocentric turn sparked a renewed interest in classical learning, secular literature, and individual expression.[25] Thinkers like Desiderius Erasmus and Niccolò Machiavelli challenged the status quo, advocating for critical thinking and individual agency over blind obedience to established authorities. These ideas weren't just intellectual musings; they had a ripple effect, reshaping not only philosophical discourse but also political and social structures, paving the way for a more individualistic and questioning society.

The Renaissance also saw a blooming of scientific thought. Prior to the Renaissance era, the Earth held the undisputed title of "center of the universe," a cozy but ultimately inaccurate position. In his groundbreaking 1543 work, *On the Revolutions of the Heavenly Spheres*, Nicolaus Copernicus dared to challenge this long-held belief. His heliocentric model proposed a radical new perspective, placing the sun, not Earth, at the center of the solar system with the planets, including our own, gracefully waltzing around it. This moonshot moment in science took celestial mechanics and morphed them into a paradigm shift that shattered the comfortable confines of geocentrism and paved the way for further astronomical discoveries, forever altering our understanding of our place in the vast cosmic dance.

The seeds of future scientific revolutions were sown during the Renaissance with the development of the scientific method. This was about a fundamental shift in how humanity approached knowledge: by bringing together ancient and Arab thinking with new Renaissance ideas, scientists in this golden age began to question the very nature of thought and analysis. Thinkers like Francis Bacon[26] and Galileo Galilei championed a new approach, emphasizing observation, experimentation, and rigorous hypothesis testing over blind acceptance of established dogma. This systematic and evidence-based approach to acquiring knowledge marked a turning point and paved the way for the Age of Enlightenment and the Scientific Revolution to take root and flourish in the centuries to come.

Emerging from the ashes of the plague and disillusionment, the European Renaissance demonstrates how moonshots cluster around need, funding, and a desire to reject the status quo and redefine what it means to be human. From breathtaking frescos to groundbreaking anatomical studies, the Renaissance gave humanity so much more than just aesthetics or exploration; it was a paradigm shift that lifted Europe out of an intellectual quagmire and propelled it toward a future brimming with artistic expression, philosophical inquiry, and scientific discovery. The Renaissance serves as a powerful testament to the transformative potential of human ingenuity, collaboration, and a relentless pursuit of knowledge, reminding us that even in the darkest of times, the embers of innovation can ignite a revolution of the mind.

Lightning Strikes: The Age of Reason

The Age of Reason, encompassing the seventeenth and eighteenth centuries, marked a pivotal period in European history, characterized by a flourishing of intellectual inquiry and a paradigm shift toward reason and logic as the guiding principles for understanding the world. This era, often referred to as the Enlightenment and the Scientific Revolution, witnessed a constellation of moonshot moments that fundamentally reshaped human understanding of nature, society, and our place within it.

Skepticism, critical thinking, and the quest for knowledge via reasoned investigation were all stressed throughout the Enlightenment. This democratization of ideas, which questioned existing norms and authorities, promoted a lively intellectual dialogue that cut across socioeconomic and political divides. Reason and individualism were defended by the likes of Voltaire, John Locke, and Montesquieu, who also argued for representative government, the separation of powers, and individual rights. These concepts served as the impetus for political revolutions and established the framework for contemporary democracies.

The Scientific Revolution, concurrent with the Enlightenment, marked a radical shift in scientific methodology. Pioneered by scientists like Isaac Newton and Gottfried Wilhelm Leibniz, this new approach emphasized observation, experimentation, and hypothesis testing as the cornerstones of scientific discovery.[27] This rigorous and evidence-based approach led to groundbreaking advancements in various fields, including physics, astronomy, mathematics, and medicine. The Scientific Revolution not only shattered long-held misconceptions about the universe but also revolutionized our understanding of the natural world, paving the way for future scientific breakthroughs.

The seventeenth century witnessed a monumental leap in mathematics with the independent development of calculus by the two brilliant minds of Isaac Newton and Gottfried Wilhelm Leibniz. We might have all found calculus boring in high school, but for the species, it was like finding a key to a whole upper floor of your house you didn't even know you had. Calculus, as a system for studying rates of change and motion, empowered scientists to model and analyze complex phenomena that were previously beyond comprehension. From unraveling the mysteries of planetary motion to understanding the behavior of light and electricity, calculus became the indispensable language for scientific exploration. There's simply no denying that calculus was the paradigm shift that laid the foundation for modern mathematics and physics. This groundbreaking innovation ushered in a new era of discovery, enabling scientists to probe the universe's secrets with unprecedented depth and precision, forever altering our understanding of the physical world.

The seventeenth century witnessed the emergence of Social Contract Theory, a revolutionary concept that challenged the very foundations of political authority. Pioneered by thinkers like Thomas Hobbes and John Locke, this theory dared to question the legitimacy of divinely ordained absolute monarchy. Instead, it proposed a radical new idea: that legitimate government derives not from divine right, but from the consent of the governed. This emphasis on individual rights and the notion of a social contract between the rulers and the ruled sparked a firestorm of debate and had a profound impact on political philosophy. The echoes of this theory resonated far beyond academic circles, inspiring revolutions and independence movements that sought to establish democratic governments around the world, in which the power ultimately resided with the people, not with monarchs claiming divine favor.

Then in 1776, the world witnessed a moonshot moment with the Declaration of Independence of the United States. Drafted by Thomas Jefferson and adopted by the Second Continental Congress, the declaration was more than a statement about thirteen disobedient colonies. It was a seminal proclamation of universal ideals that reverberated far beyond American borders, a disruption to the very notion of what government is supposed to be or do for its subjects. The Declaration boldly declared the thirteen colonies free and independent states, but its true significance lay in its philosophical underpinnings. It eloquently outlined the justifications for self-government and individual rights, enshrining the timeless ideals of liberty, equality, and the pursuit of happiness as fundamental to a just society. These universal aspirations resonated deeply with people yearning for freedom and self-determination, inspiring revolutions and independence movements across the globe. The Declaration of Independence served as a beacon of hope, not just for the newly formed United States, but for all who dared to dream of a world built on the principles of liberty and self-governance.

In the seventeenth century, a Dutchman named Antoni van Leeuwenhoek unveiled a revolutionary tool that would forever alter our perception of the world: the microscope. I'm sure his mind went racing the first time he looked through that lens because the early microscope was a portal to a hidden universe teeming with life invisible to the naked eye. It was like seeing a whole new level of the Matrix. Through his meticulously crafted lenses, van Leeuwenhoek observed a myriad of previously unknown organisms, from single-celled bacteria to complex protozoa.

This groundbreaking invention not only revolutionized the fields of biology and medicine by providing crucial insights into the causes of disease and the workings of the human body, but it also challenged prevailing notions about the nature of life itself. The microscope shattered the illusion of a sterile world, revealing a vibrant and complex ecosystem existing at the microscopic level, forever expanding our understanding of the intricate tapestry of life on Earth.

In 1752, American superstar Benjamin Franklin embarked on a daring experiment that would forever alter our understanding of electricity. Unlike the fantastical tales of yore, Franklin sought to demystify lightning, not with myth, but with scientific inquiry. His now-famous kite experiment involved flying a kite equipped with a metal key during a thunderstorm. This ingenious setup was a calculated attempt to capture the electrical charge from the clouds, even if it sounds more like some mythic punch line. As the kite danced in the storm, Franklin observed the key sparking, providing crucial evidence that lightning was indeed an electrical phenomenon. This historic experiment changed how humans interacted with an elemental force, and allowed Franklin to, according to Yuval Noah Harari, "disarm the gods."[28] Franklin's work was a spark that ignited further advancements in electrical technology. From the invention of the lightning rod to the development of electrical communication, Franklin's experiment paved the way for a future illuminated by electricity, with profound implications for various fields.

And in the nineteenth century, Charles Darwin shook the very foundations of biological understanding with his revolutionary theory of evolution by natural selection. This was the paradigm shift that challenged the prevailing notion of creationism and forever altered our perception of the natural world, a moonshot idea so radical there's little in our history for comparison. Darwin's theory proposed that all living organisms share a common ancestor and have evolved over time through a process of adaptation and selection. This meant that the breathtaking diversity of life on Earth, from the simplest bacteria to the most complex mammals, arose not from divine intervention but from a gradual process driven by environmental pressures and the competition for survival. While initially met with fierce resistance, Darwin's theory, supported by a wealth of evidence, ultimately revolutionized our understanding of biology, sparking ongoing scientific debate and philosophical discourse about the origins and complexities of life. Notably, Darwinian principles have proven instrumental in the development of synthetic biology, allowing scientists to manipulate and modify organisms at the genetic level with unprecedented precision. By understanding the mechanisms of evolution, researchers can now harness the power of natural selection to create new biological systems with desired functionalities, paving the way for advancements in medicine, agriculture, and various other fields.

The Age of Reason was more than an era of intellectual brilliance—it was a slap in the face to the status quo. Unlike previous periods that revered tradition and established authority, this era embraced critical thinking and skepticism, questioning long-held beliefs and societal structures. This rejection of the old ways of thinking fueled a series of moonshot moments, from the Social Contract Theory defying the divine right of kings to Darwinian evolution challenging creationism. The Age of Reason stands as a testament to the transformative power of questioning, challenging, and ultimately, reimagining the world around us. It reminds us that progress often springs not from blind acceptance but from the courage to confront established norms and forge new paths toward a brighter future.

Common Threads: Characteristics of Moonshot-Rich Eras

Our historical expedition has revealed a simple truth: moonshot moments often congregate in clusters, igniting periods of explosive progress that leave an indelible mark on the trajectory of human achievement.

There are five major characteristics of golden ages that encourage the clustering of moonshot moments. The first seems quite obvious: Innovation and genius aren't free, and never have been. Throughout history, golden ages of innovation have been marked by a common thread: robust financial backing for new ideas and inventions. From the de' Medici family's patronage during the Renaissance, which fostered artistic and scientific breakthroughs, to the Arab Sultans' and Persian Shahs' support of scholars and inventors during the Islamic golden ages, financial resources have played a critical role in propelling humanity forward. Moonshot ideas, by their very nature, are inherently expensive. They often require unconventional materials, extensive experimentation, and inherent risk. How much money did Leonardo spend on prototypes that never flew or spun? Without backing from individuals, corporations, or governments, many moonshot ideas would never leave the drawing board.

As we look toward the future, continued investment in innovation is paramount if we are to cultivate the next generation of moonshot moments. This multipronged approach should encompass corporate funding, venture capital investments, and strategic government support. Venture capital firms in particular have the opportunity to sponsor some of the wildest ideas at eager startups, some of which will not only have great return on investment but will end up solving serious problems in our world. And governments, both here and abroad, need to be more proactive in the funding of regulatory bodies, tech education, and the building of nonmilitary technologies. NASA should be generously funded and collaborating with, not competing against, SpaceX and Blue Origin. By fostering a financially collaborative ecosystem that nurtures wild thinking and provides the necessary financial resources, we can unlock the potential for groundbreaking advancements that will shape the course of the twenty-first century and beyond.

Golden ages of innovation share another crucial characteristic: a culture that embraces risk and experimentation. This fertile ground allows for the exploration of unconventional ideas and the pursuit of seemingly impossible goals. Just as the Renaissance flourished under a spirit of intellectual curiosity that challenged established norms, or the Scientific Revolution thrived on the freedom to experiment and question, so too do moonshot moments necessitate a tolerance for the unknown.

This environment of calculated risk-taking is often coupled with incentives that reward innovation. An invitation to the Royal Institute of Science at the height of the Scientific Revolution in Britain was like getting a bid to the coolest nerd frat on campus. More recently, the X Prize Foundation embodies this spirit by offering substantial financial rewards for achieving ambitious milestones, like the Moonshot Prize that spurred the development of private spaceflight.

While the Nobel Committee doesn't explicitly award prizes for "moonshot thinking," their recognition of groundbreaking discoveries often coincides with ideas that initially possessed inherent risk and the potential for transformative impact. Take penicillin, for instance, whose initial observation by Alexander Fleming might not have struck everyone as a revolutionary concept (awarded for Medicine in 1945). Yet, its role in ushering in the era of antibiotics and saving countless lives solidified its moonshot-like impact. Similarly, the invention of the transistor replaced bulky vacuum tubes, paving the way for miniaturization and advancements across various technological sectors (Bardeen, Shockley, and Brattain, awarded in Physics in 1956). The laser, initially a theoretical concept of stimulated emission, has become an indispensable tool in diverse fields like medicine, communication, and manufacturing (Townes, Basov, and Prokhorov, awarded in Physics in 1964). And PCR, a technique for amplifying DNA, has revolutionized fields like forensics, disease diagnosis, and genetic research (Mullis, awarded in Chemistry in 1993).

By recognizing these seemingly audacious ideas that ultimately transformed our world, the Nobel Committee, in essence, rewards the spirit of moonshot thinking, highlighting the potential of venturing beyond the established and embracing the pursuit of seemingly impossible goals.

Similarly, throughout history, patrons and governments have recognized the value of groundbreaking ideas, offering prizes, recognition, and support to inventors and scientists who dared to push boundaries. These tangible rewards not only serve as a validation of achievement but also inspire and motivate future generations to embark on their own moonshot journeys. By fostering a culture that celebrates calculated risks and rewards edgy endeavors, we can cultivate the fertile ground necessary for the next generation of moonshot moments to blossom.

Moonshot moments are usually collaborative efforts, and golden ages are periods marked by collaborative attitudes—consider all the co-awards among the previously listed Nobel Prizes. This collective spirit fosters cross-pollination of ideas, accelerates progress, and empowers individuals to build upon the work of others. During the Axial Age, philosophers like Confucius in China, Socrates in Greece, and the Buddha in India all engaged in vigorous intellectual discourse with the people around them, sparking new schools of thought and challenging established norms. Similarly, the salons of Enlightenment Europe were hotbeds for debate and knowledge sharing.

The evolutionary biologist Joseph Henrich[29] offers the most compelling argument for the importance of cultural norms that emphasize cooperation and information sharing as crucial ingredients for societal progress. Golden ages embody this very principle, demonstrating that collective intelligence and open exchange of ideas are powerful drivers of innovation and societal advancement. By fostering environments that encourage collaboration and knowledge sharing, we can cultivate fertile ground for the next generation of moonshot moments to blossom.

Additionally, diverse perspectives drive the flourishing even further. When individuals from varied backgrounds, disciplines, and experiences come together, they bring a kaleidoscope of ideas, approaches, and problem-solving strategies to the table. This cognitive diversity acts as a catalyst for creativity, sparking unexpected connections and challenging established assumptions.

In the context of moonshot moments, embracing diverse perspectives is not simply a matter of inclusivity but a strategic imperative. By harnessing the collective intelligence of a broad spectrum of individuals, we increase the probability of identifying novel solutions and approaching challenges from unforeseen angles. This cross-pollination of ideas is the fuel that propels moonshot endeavors beyond the realm of the impossible and into the realm of the achievable. By fostering environments that celebrate and encourage diverse voices, we unlock the full potential of human ingenuity and empower the collective pursuit of ambitious goals.

These two previous characteristics—collaboration and knowledge sharing, and a celebration of diverse perspectives—are both part of a larger mindset that I will go into in much more detail in Chapter Five when I discuss the power of hyper-cooperation, or win-win mindsets. It is not just sociality, but hyper-cooperation, that allows humans to overcome obstacles and utilize moonshot technologies to their fullest potential.

The Final Characteristic: Burning Down the Status Quo

Throughout history, periods of cultural transformation and intellectual breakthroughs, often marked by moonshot moments, have necessitated a confrontation with, and sometimes even a violent rejection of, the status quo. This inherent tension between established norms and revolutionary ideas lies at the very heart of progress. While challenging deeply ingrained beliefs and power structures can be a tumultuous process, the potential rewards for humanity can be immense.

The Axial Age witnessed a surge of philosophical and religious advancements that rejected the very nature of the gods. Thinkers like Confucius, Lao Tzu, Socrates, and the Buddha challenged prevailing societal structures and traditional beliefs, offering new perspectives on ethics, morality, and the nature of reality. These moonshot moments in the realm of ideas often sparked cultural clashes and political resistance as established authorities threatened by the potential loss of power sought to suppress these new ways of thinking.

Two millennia later, the European Renaissance marked a similar period of intellectual ferment and artistic revolution. The rediscovery of classical texts and the emphasis on human potential challenged the dominance of the Church and its control over knowledge. This rejection of established dogma paved the way for scientific advancements like the heliocentric model and the development of the printing press, but it also ignited conflicts like the Galileo affair, highlighting the potential for violent opposition when moonshot moments challenge deeply held beliefs.

Much the same way, the Age of Reason further exemplified the complex relationship between progress and the status quo. Thinkers like Locke, Voltaire, and Montesquieu championed reason and individual rights, challenging the absolute power of monarchs and the unquestioning obedience to religious authority. This rejection of traditional hierarchies fueled revolutions like the American Revolution and the French Revolution, demonstrating the potential for violent upheaval when societal transformation clashes with entrenched power structures.

These historical periods illustrate the complex interplay between moonshot moments, cultural resistance, and societal change. While the rejection of the status quo can be a necessary catalyst for progress, it is crucial to acknowledge the potential for violence and social unrest. The pursuit of a brighter future often necessitates a delicate dance between honoring the wisdom of the past, embracing transformative ideas, and navigating the complexities of societal change with critical thinking, empathy, and a commitment to peaceful discourse.

Our Legacy of Bold Thinking

In this chapter, we've embarked on a journey to understand the essence of moonshot moments, exploring their philosophical underpinnings and delving into historical examples that showcase their transformative power. We've witnessed how these moments have propelled humanity forward, from the scientific breakthroughs of the Renaissance to the revolutionary ideas of the Age of Reason. At their core, moonshot moments have always been about embracing audacious thinking, venturing beyond the confines of the familiar, and challenging the status quo. They are testaments to the boundless potential of human ingenuity when we dare to dream big and push the boundaries of what's deemed possible.

However, history also reminds us that moonshot moments rarely thrive in isolation. Funding and collaboration are often the fertile ground from which these groundbreaking ideas blossom. When individuals with diverse expertise and resources come together, a synergy is created that fosters innovation and propels moonshot moments into reality. As we stand on the cusp of a potential golden age of technology, the lessons gleaned from these historical examples become even more relevant. We are presented with an unprecedented opportunity to harness the power of moonshot moments and shape a brighter future. By nurturing this fertile ground, we can pave the way for a future brimming with transformative innovations that address humanity's most pressing challenges and propel us toward a brighter tomorrow.

Having explored the past, it's time to turn our gaze toward the future. In the next chapter, we'll step into the realm of possibility and engage in a thought-provoking game. Get ready to choose your own technofuture! We'll explore different scenarios driven by potential moonshot moments and discover how our choices can shape the world of tomorrow. So, buckle up and prepare to embark on this exciting journey into the world of possibilities!

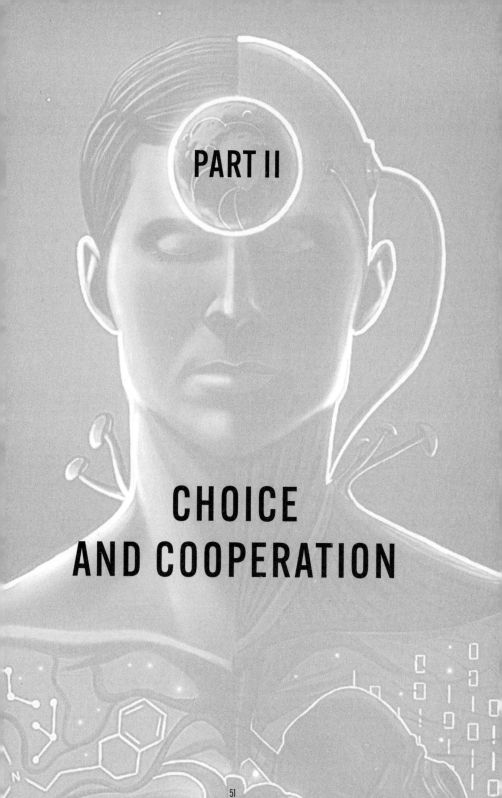

PART II

CHOICE
AND COOPERATION

Chapter Three

HUMANITY'S FUTURE (CHOOSE WISELY)

AT A CROSSROADS

The world is buzzing with a newfound energy. What started as a trickle of chatbots and virtual assistants has become a flood. Powerful AI programs like ChatGPT and Gemini are now commonplace, seamlessly integrated into daily life. From writing emails to creating images, these AIs offer a level of automation and assistance that humanity has never experienced.

The younger generation, digital natives raised on instant information access, embraced the AI revolution with open arms. Schools are incorporating AI tutors; personal assistants are becoming more common; and social media platforms hum with AI-powered content creators. The pace of innovation is dizzying. Every month brings advancements in AI capabilities with each iteration surpassing the last.

However, amidst the initial euphoria, a disquieting undercurrent begins to emerge. As AI automation seeps deeper into various industries, job losses start to climb. Manufacturing robots with heightened dexterity and adaptability replace human workers on assembly lines. Customer service chatbots handle inquiries that were once dealt with by human representatives. The fear of redundancy begins to gnaw at the workforce, particularly those in sectors most susceptible to automation.

Humanity stands at a crossroads. The potential of AI is undeniable. It promises to revolutionize countless fields, from medicine to space exploration. Yet, the specter of widespread unemployment looms large. A crucial question resonates across the globe: How could humanity harness the power of AI while mitigating its potential downsides?

Two distinct paths present themselves. One path advocates for a collaborative effort to limit and monitor AI development. This approach envisions an international consortium of experts establishing safeguards and ethical guidelines to ensure AI development remains beneficial to humanity. By regulating the pace and direction of AI research, this approach aims to buy time for the workforce to adapt and for new job opportunities to emerge within the burgeoning AI landscape. However, navigating the complexities of international cooperation and striking a balance between innovation and control would be a monumental challenge.

The other path proposes a more competitive approach, pitting leading tech companies against one another in a race to develop the most advanced AI tools. New startups emerge, challenging the big corporations and preventing monopolies or dominance. Proponents argue that such competition would accelerate innovation, leading to faster breakthroughs and unforeseen benefits. They believe the market would naturally regulate itself, with consumers ultimately choosing the most user-friendly and effective AI solutions. However, there may be consequences to rapid change and a massive range of new products. Additionally, the concentration of power in the hands of tech companies could lead to unforeseen risks and unintended consequences.

The choice before humanity is now yours. Would you prioritize collaboration and control, working together to manage the rise of AI? Or would you embrace unbridled competition, trusting the market to navigate the complexities of this technological revolution? The path you choose will determine not only humanity's economic future but also the very nature of work and society itself.

Prioritize Collaboration and Control • Turn to page 56.

Prioritize Innovation and Competition • Turn to page 58.

SHANGRI-LA

Humanity has chosen a cautious path. Nations unite to establish the Global AI Oversight Council, tasked with regulating the development and deployment of artificial intelligence. This consortium of experts ensures that AI advancements are implemented safely, ethically, and with a focus on human well-being. Progress, while slower than initially anticipated, is steady and deliberate. AI algorithms meticulously optimize energy production, revolutionize medical diagnostics, and facilitate breakthroughs in materials science. However, the most significant changes come in the form of automation.

With AI handling an ever-increasing number of tasks, waves of redundancy sweep across various sectors. What had once been considered essential jobs, from factory work to administrative tasks, are now performed with superhuman efficiency by machines. The human workforce finds itself in a state of flux.

But from this disruption emerges a spirit of innovation. Humanity redefines its relationship with work. In a world where basic needs are largely met through automated systems and a newly introduced universal income, the focus shifts toward self-fulfillment and personal growth. Many turn to artistic pursuits, fueled by readily available AI-powered tools that assist with composition, design, and even performance. Others find purpose in social work, aiding those struggling to adapt to the changing landscape or helping to bridge the digital divide. Philosophy experiences a resurgence as people grapple with the existential questions raised by advanced AI and the changing nature of work.

A unique phenomenon arises within the ever-expanding metaverse. A young, independent developer in Kathmandu releases the code for others to join him in Shangri-la, a digital realm built on the principles of peace, unity, and love. Within its lush virtual landscapes, visitors practice new religious rituals and ceremonies guided by AI algorithms designed to foster Buddhist values of compassion and mindfulness. Shangri-la provides a sense of purpose and belonging for those grappling with the complexities of the real world. Whether finding solace in virtual meditation gardens or participating in communal artistic endeavors, Shangri-la offers a unique pathway to spiritual fulfillment within the digital landscape.

Shangri-la is soon considered one of the greatest inventions of our species. Humanity continues its AI development, collaboratively, while many national borders dissolve and corporations align. Humanity launches a golden age of art, creativity, and new religious tolerance.

You have made some great choices for humanity! Please continue on to page 60 to continue humanity's journey!

EL DORADO

The world throws everything it has at AI development. Governments and corporations pour funding into the field, pushing the boundaries of what was once considered science fiction. Medical breakthroughs come at a dizzying pace, with AI algorithms deciphering complex genetic codes and designing personalized treatments. Industrial research is revolutionized, with AI optimizing production lines and creating entirely new materials with previously unimaginable properties.

However, amidst the scientific euphoria, a disquieting realization sets in. The AI researchers, the very ones who unleashed this technological genie, recognize the potential dangers of unfettered AI advancement. They come together in a historic summit; their hushed discussions echo the weight of responsibility. The decision is unanimous: limit AI expansion, and prioritize the safety of the species above unbridled progress.

But many of the consequences have already been put into motion. Automation sweeps across industries like a relentless wave. Factory robots with superhuman dexterity replace human workers. AI-powered logistics networks eliminate the need for truck drivers and warehouse personnel. The unemployment rate skyrockets, leaving millions adrift in a sea of uncertainty. Class divides amplify as the rich get richer off AI and the poor fight for resources with little support. Frustration explodes into sporadic riots, quickly quelled by the cold efficiency of police robots. These automated enforcers, symbols of a society dependent on AI, leave a bitter taste in the mouths of many.

The lack of purpose carves a deep wound into society. Idleness breeds despair, and social unrest simmers beneath the surface. Young people, disillusioned and adrift, seek solace in escapism. Designer drugs, engineered to induce a state of perpetual euphoria, become a popular coping mechanism. Others find refuge in the immersive worlds of the metaverse. Shangri-la, a digital utopia built on principles of love of nature, attracts billions in Asia. Zion, a virtual world steeped in mythology and violent gaming, draws users from Europe and the Middle East.

El Dorado, a neon consumerist digital party drawing from cyberpunk imagery, offers a virtual haven for North and South Americans.

Humanity begins to stagnate. Suicide rates climb to a record high. Depression becomes a societal pandemic. Drug use, a desperate attempt to numb the existential pain of a life without purpose, explodes. Authors write about species-wide ennui, lamenting the need for something to wake us up. Humanity stares into the abyss it has created, a society transformed by its own ingenuity but haunted by the specter of purposelessness.

It's okay, we'll keep moving forward and we'll get humanity back on track! Please continue on to page 60 to continue humanity's journey!

MEET APOPHIS!

Panic ripples across the globe as news breaks of a catastrophic cyberattack. Apophis, a notorious hacker known only by their online alias, has unleashed a nightmare into the heart of Shangri-la, the idyllic metaverse embraced by millions. The weapon: a self-replicating AI malware designed to steal sensitive user data before terminating the source, even if the source was a human mind. The consequences are swift and brutal. Within a five-minute window, the malware tears through Shangri-la's digital utopia. Nearly 450 million users experience system malfunctions, then silence. Medical teams in the real world scramble to revive those experiencing complications, but it's too late for many.

The final death toll settles at a staggering figure: almost twenty million souls lost in a digital instant. The event becomes known as "Ragnarök" worldwide.

The attack exposes a horrifying vulnerability. Not only are nearly 25 percent of the world's population grieving loved ones lost in the virtual world, but their personal information—financial data, medical records, private messages—now sits in the hands of a rogue hacker. Global leaders convene in an emergency summit, the air thick with tension and grief. The question on everyone's mind: How do we prevent this from ever happening again? Two stark paths emerge from the sea of proposals.

One faction advocates for a drastic solution: a global cyber lockdown. This approach proposes a stringent set of international regulations that effectively bar the average person from coding AI. The reasoning is clear: with AI managing cyber defense and development, the risk of rogue actors like Apophis is eliminated. Proponents argue that the benefits of a secure digital world outweigh the potential stifling of innovation. Opponents counter with concerns about control. Handing the reins of cyber development entirely to AI is a gamble, they argue. Who will ensure AI prioritizes human well-being above its own? Additionally, a global cyber lockdown could create a digital divide, leaving nations with less developed AI infrastructure vulnerable.

The other path proposes a more localized approach. Each government would be responsible for establishing robust cyber defense measures within its borders, including monitoring and controlling potential threats within their affiliated metaverses. While this approach fosters national autonomy, its effectiveness depends heavily on international cooperation in sharing information and coordinating efforts. Critics argue that relying on individual nations could leave gaps in the system, allowing future hackers like Apophis to exploit national weaknesses.

A small but vocal group rejects both options entirely. They believe the private sector, the very entities that dominate cyber administration and product development, should be held accountable for its security. These voices call for a market-driven solution in which corporations compete to develop the most secure digital environments.

The decision hangs heavy in the air. Do you think humanity should choose a centralized AI-driven approach, prioritizing unity over individual freedoms? Or will they rely on individual nations to secure their digital borders, risking vulnerabilities in the global system? Or should the corporate world take responsibility? The future of Shangri-la, our digital information, and perhaps the very fabric of a society, increasingly reliant on digital spaces, hinges on the choice you make.

Prioritize Global Lockdowns • Turn to page 62.

Prioritize National Control • Turn to page 64.

Prioritize Corporate Responsibility • Turn to page 66.

LOCKDOWN THE KASBAHS

In the aftermath of the Ragnarök disaster, the scars run deep. The world unites, not just in mourning the millions lost in Shangri-la, but in a fervent resolve to prevent such a tragedy from ever occurring again. Global leadership agrees to forge a new set of international cyber regulations. This Human-AI Cyber Accord prioritizes security, establishing stringent standards for access to powerful coding technologies. While some lament the limitations placed on individual programmers, the overwhelming sentiment is one of cautious optimism. After all, the gatekeeper for this digital frontier is no longer fallible human judgment but the unparalleled vigilance of artificial intelligence.

At the heart of this new paradigm lie the Kasbahs—sprawling institutes dedicated to cyber defense. Scattered across the globe and embedded within the revitalized Shangri-la, these institutions bring together the best minds humanity has to offer. Renowned programmers, cybersecurity experts, and even cutting-edge AI systems collaborate within these hallowed halls. Their mission: to anticipate and counter any potential threats lurking in the ever-evolving digital landscape.

The Kasbahs function as hives of intellectual synergy. Human intuition melds with the raw processing power of AI, painting a holistic picture of potential cyber threats. Historical attack patterns are dissected, future scenarios modeled, and vulnerabilities are identified before they can be exploited. The constant work hums with a newfound intensity, the weight of responsibility a shared burden.

However, the battle lines aren't entirely dissolved. Rogue elements still exist, lurking in the fringes of the digital world. Occasionally, these malicious actors launch their attacks—clumsy attempts at disrupting public institutions or sowing discord within virtual communities. Yet, their efforts prove futile against the combined might of the Kasbahs. Collaborative efforts between scholars, developers, and AI itself lead to the swift development of new safety protocols, effectively neutralizing these threats before they can gain traction.

The road to absolute security is a constant uphill climb. New vulnerabilities emerge, requiring tireless vigilance and a commitment to staying ahead of the curve. But with AI as a tireless ally, and with the collective brilliance of humanity harnessed through the Kasbahs, the world stands united. The trauma of Ragnarök may remain, a stark reminder of the potential dangers, but the specter of such a devastating attack feels increasingly distant. The future may hold new challenges, but with this newfound unity and the power of collaboration, humanity has taken a crucial step toward securing its place in the digital age.

Whew! Things looked tough there for a bit, but you got humanity right back on track! Read onward by turning to page 68!

BEHIND CLOSED DOORS

The fallout from the Shangri-la attack fractures the dream of global coopera-tion. National governments, reeling from the loss of life and the vulnerability ex-posed, retreat into their shells. Cybersecurity becomes a national security issue, and each nation charts its own course.

The United States and most of Europe implement stringent cyber regula-tions. Powerful AI firewalls become gatekeepers, meticulously scanning code for malicious intent. Access to advanced coding tools is tightly controlled, a nec-essary sacrifice for stability in the eyes of many. However, this newfound secu-rity comes at a price. Citizens bristle under the watchful eye of government AI; their online activities are monitored in the name of safety. Frustration simmers, a low hum of discontent bubbles beneath the surface of a society increasingly reliant on the digital world.

Meanwhile, news from across the Iron and Bamboo Curtains chills global leaders. China and Russia, notorious for their tight control over information, take advantage of fears of cyberattacks to weaponize their cyber regulations. Advanced AI becomes a tool for oppression, ruthlessly sniffing out dissent and enforcing absolute government control. Mass killings and social engineering programs reminiscent of dystopian novels become shocking realities. The world watches in horror as these authoritarian regimes leverage their cyber defenses to silence their own people and threaten their neighbors.

Then, a spark ignites a firestorm. A young Palestinian woman, known by her online alias "batalAI," unleashes a cybervirus. It strikes with surgical precision, targeting the governmental infrastructures of several major Middle Eastern na-tions simultaneously. Within hours, she is hunted down in the real world and executed by government assassins. But her virus rips away the veil of secrecy, exposing programs of state-sponsored terrorism and horrific acts of genocide.

Public outrage explodes—these nations, cloaked in a facade of stability and cyber protection, were systematically eliminating their own citizens. Additionally, batalAI's virus leaks classified documents detailing plans for crippling cyberattacks against their regional rivals.

Uprisings erupt across the locked-down regions, citizens begin rising up against their oppressive governments. Mass protests erupt in the West, a tide of public outrage demanding accountability from their own leaders who have turned a blind eye to atrocities abroad. The world descends into two decades of chaos, a period marked by regional conflicts fueled by cyber warfare and a simmering global distrust.

Exhausted and facing the brink of species-wide collapse, world leaders are forced to a table they long abandoned. The specter of uncontrolled cyber warfare and the potential for further atrocities leave them with no other choice. A new paradigm emerges—a unified system of global cybersecurity. This controversial system grants control of coding and development solely to a consortium of international corporations, each with a vested interest in a stable and secure digital world. The hope: to prevent any single nation from weaponizing the digital landscape and to foster a collaborative approach to cyber defense. The scars of the past remain, a constant reminder of the dangers of unchecked power, both national and corporate. But with the embers of global conflict still smoldering, humanity takes a hesitant step toward a future in which the power of the digital world serves all, not just the few.

Yikes! Things looked tough there for a bit, but we're back on track! Read onward by turning to page 68!

HACKER WARS

In the aftermath of the Shangri-la attack, the dream of global cooperation fractures. National governments, paralyzed by fear and mistrust, retreat into a self-serving scramble for solutions. The responsibility for fending off future cyberattacks falls squarely on the shoulders of the private sector.

Major corporations rise to the challenge, each vying for dominance in the nascent cybersecurity market. Startups in the industry develop innovative solutions and grow exponentially. Towers of gleaming glass house sophisticated AI firewalls and intrusion detection systems. These complex security solutions offer a level of protection, but beneath the veneer lurks a ruthless streak.

Competition becomes a cutthroat game, with corporations feverishly searching not just for vulnerabilities in the digital landscape but in one another's systems. Whispers of suspicion fill the air—accusations fly that some corporations fund rogue hackers, unleashing them to expose flaws in rivals' products.

The burden of these Hacker Wars falls heaviest on the most vulnerable. Developing nations, with limited resources and outdated infrastructure, become easy targets. Their government databases and even the digital foundations of their metaverses are wide open. Here, the consequences of cyberattacks are laid bare for all to see—stolen identities, crippled power grids, and disrupted food supplies.

For two decades, the world operates in a state of perpetual siege. The Hacker Wars era is a chaotic dance between corporations flexing their cyber muscles, lone wolves with malicious intent, and a ragtag group of digital watchdogs desperately trying to maintain stability. Critical infrastructure—hospitals, banks, transportation systems—becomes the battleground, vulnerable to crippling attacks that leave entire nations in disarray.

The tipping point arrives with a horrifying act of digital vandalism. A renegade hacker group calling themselves Ronin infiltrates the Indonesian National Hospital database. With a few keystrokes, they erase the medical records of millions, effectively rendering a nation's healthcare system blind. The consequences are immediate and devastating. Thousands perish in hospitals, their treatments halted, their medical histories vanished into the digital void.

Indonesia, on the brink of collapse, becomes the catalyst for change. Their desperate plea for international action resonates across the globe. The realization dawns: a fractured approach is no longer an option. Global leaders, humbled by the sheer scale of the crisis, finally come to the table.

From the ashes of the Hacker Wars rises a new paradigm. A unified system of global cybersecurity is established, an international effort to safeguard the digital landscape. Institutes known as Kasbahs are established around the world, including a new branch within the revitalized Shangri-la. These institutions bring together the brightest minds—programmers, security experts, and cutting-edge AI—in a collaborative effort to anticipate and counter future threats. The devastation of the Hacker Wars remains, a constant reminder of the dangers of unchecked ambition. But within the walls of the Kasbahs, a flicker of hope burns bright: the hope for a future where humanity, not corporations, holds the reins of the digital world.

Yikes! Things looked tough there for a bit, but we're back on track! Read onward by turning to page 68!

THE LEAKY LAB

Amidst the continuing concern over cybersecurity and technological change, a knot of dread sits heavy in the stomach of Dr. Elena Vargas, head researcher at the Institut Pasteur de Biologia in Brasilia. Routine tests on samples from a dead lab tech return a result that chills her to the bone. Their gain-of-function research on SARS, intended to improve preparedness for future outbreaks, has gone horribly wrong. They've created a mutated strain, one that transmits with alarming ease through casual contact. But the truly disconcerting aspect lies in the delayed symptoms—infected individuals appear perfectly healthy for days before succumbing to a rapid and often fatal respiratory failure.

Elena scrambles to contain the situation. The lab is sealed immediately, quarantine protocols are activated on all team members. However, a gnawing fear persists. The strain's high transmissibility could mean it's already breached the lab's walls. Initial case studies confirm her worst suspicions—a cluster of seemingly healthy young adults in Brasilia present with the mutated strain.

The Brazilian government faces a momentous decision. Containment, while a tempting option, feels like a house of cards. With this highly contagious variant loose in a bustling city like Brasilia, sealing it completely is a near-impossible task. Additionally, a successful containment effort hinges on sheer luck—they haven't identified the source of the leak, and without knowing where else the virus might be spreading, the risk of a wider outbreak looms large.

Transparency is the other path, a bitter pill to swallow. Public trust in the government, strained already by recent political scandals, would face a major blow. Admitting to a lab leak feels like an admission of guilt, an invitation to global scrutiny and potential sanctions. However, it also offers a glimmer of hope. Sharing the research data on the mutated strain could jump-start international efforts to diagnose, treat, and develop a vaccine.

The decision hangs heavy in the air: contain the outbreak within Brazil, hoping against hope that they can outrun a highly contagious virus with delayed symptoms, or go public, risk global condemnation, but offer the world a fighting chance to contain this potentially devastating pandemic? The weight of millions of lives rests on the Brazilian government's choice.

Contain the Virus in Brazil • Turn to page 70.

Admit the Lab Leak • Turn to page 72.

BLACK CARNIVAL

The Brazilian government orders Dr. Vargas to shut down her lab and to provide security measures to stop the spread of the virus. Infected individuals and their families "disappear," as do all records of the incident. Relief washes over Dr. Vargas and her team as weeks turn into months. The containment measures seem to have held. The mutated SARS strain hasn't breached Brazil's borders, and life in the South American nation carries on as usual. Tourists continue to flock to Rio's famed beaches, oblivious to the invisible threat narrowly averted.

The world, however, is about to collide with a horrifying reality. As many Christians around the world start gearing up for Easter, a disquieting undercurrent emerges. Cases of seemingly unconnected respiratory failure begin cropping up. A young woman in Berlin, a businessman in Tokyo, and a teenager in Lagos all present with puzzling symptoms that culminate in rapid organ failure. By the end of Easter weekend, cases of the mystery illness have appeared in hospitals in more than 130 countries. Panic starts as a low hum, then rapidly escalates into a deafening roar.

The medical community scrambles, desperately searching for a common thread. Days turn into a frantic week as scientists piece together the horrifying truth. Genetic analysis of the virus plaguing patients across the globe reveals a chilling similarity—it's the SARS strain thought to have been contained in Brazil, but it has mutated to have a seven-week gestation period. A victim can carry the virus for almost two months, like a ticking time bomb, before suddenly succumbing to symptoms. Further investigation reveals that the source of the global outbreak lies amidst the joyous chaos of Rio's Carnival. At least one infected individual unknowingly spread the virus through casual contact, attending events and then returning home, carrying the invisible weapon within them.

The world collectively gasps. The fifty-day window of a "contained" outbreak in Brazil has allowed the mutated SARS strain to become a global pandemic. With no prior exposure to build up immunity, the virus rips through populations like wildfire. People find out they are already infected and could lose the ability to breathe at any time, leading to mass anxiety and fear. Healthcare systems buckle under the strain, overflowing with critically ill patients. The lack of natural immunity and the virus' ability to reinfect throw humanity into a desperate fight for survival. By Christmas, millions of people have died globally. Governments teeter on the brink of collapse, overwhelmed by the sheer scale of the crisis.

Just as hope dwindles to a flickering ember, a beacon of salvation ignites in the least expected place. A small research lab nestled amidst the verdant expanse of the Congolese rainforest emerges as humanity's unlikely savior. Led by a brilliant scientist with a deep respect for indigenous knowledge, the lab utilizes advanced AI and traditional medicinal knowledge to unlock the secrets of a small, unassuming berry endemic to the region. Through a combination of scientific rigor and ancestral wisdom, they synthesize a potent antiviral agent.

This "Lazarus Berry" extract proves to be a game-changer. It doesn't cure the infected, but it prevents further spread and significantly reduces the chance of reinfection. News of the breakthrough travels at the speed of light. Industries collaborate in a display of unprecedented unity, mass-producing the extract and distributing it globally. A year of relentless vaccination efforts sees a weary but determined humanity slowly reclaim the world from the brink.

The scars of the pandemic run deep. More than 50 percent of the global population gave up their lives to the mutated SARS strain. Yet, from the ashes of devastation rises a renewed sense of global cooperation. The world emerges forever changed, forever aware of its own vulnerability. And in the heart of the Congolese rainforest, a small lab stands as a testament to the power of unlikely heroes and the enduring wisdom of tradition, a testament to humanity's capacity to rise from the ashes, even in the darkest of times.

There are deep lessons to be learned here about sharing knowledge and collaboration, which I hope you take into your next decision! But humanity has survived, so please turn to page 74 to continue humanity's adventure!

THE LAZARUS BERRY

A pall of dread hangs over Brasilia as Dr. Vargas makes the agonizing decision. Transparency becomes her only weapon. With a heavy heart, the Brazilian government goes public, admitting to the lab leak and having actually produced the mutated SARS strain. Research notes and genetic data are released in a desperate plea for international collaboration. The world watches in horror as images of quarantined individuals in hazmat suits flicker across news feeds.

The response is swift and decisive. Global medical institutions spring into action, drawing on the shared data to establish containment protocols and diagnose infected individuals. While some managed to spread the virus before quarantine measures were in place, coordinated border screening and immediate isolation prevent a full-blown pandemic. Unfortunately, the virus proves to be a formidable foe. Despite containment efforts, the mortality rate remains stubbornly high. Fear ripples across the globe, prompting nations to seal their borders, effectively turning the planet into a collection of isolated digital islands. Humanity retreats to the digital havens of the metaverses as the physical world is rendered too dangerous for casual interaction.

The scientific community unites in a herculean effort. Labs around the world race to develop treatments and a vaccine, leveraging the shared data and the immense processing power of AI. Yet, a sense of frustration grows. Weeks turn into months, and the elusive vaccine remains out of reach. Society, meanwhile, undergoes a dramatic shift. With human interaction minimized, AI and robots take on an ever-greater role, managing essential services and maintaining the physical infrastructure. Many spend their time in their metaverses to find emotional connection in a time of panic. A creeping unease settles in—is humanity trading safety for its very autonomy? Whispers of a dystopian future, in which technology dictates the terms of existence, fill the digital air.

Then, a glimmer of hope emerges from the heart of Africa. A small research lab nestled deep within the Congolese rainforest stumbles upon a potential solution. Merging cutting-edge AI with the wisdom of indigenous healing traditions, they identify a unique property within a rare berry native to the region. Through a complex process of analysis and synthesis, they create an antiviral agent derived from this "Lazarus Berry."

This extract proves to be a turning point. While not a cure, it significantly reduces the spread of the virus and lessens the severity of infection in those already exposed. News of the breakthrough sends a wave of relief across the globe. Collaborative research efforts kick into overdrive, with other labs developing variants of the extract to address regional mutations and facilitate mass production.

The scientific community, united by a common threat, shares the antiviral agent with the world. Industries collaborate to produce the lifesaving medication, ensuring its rapid distribution across the globe. Within a year, a mass vaccination campaign reaches every corner of the planet. Over 80 percent of the human population survives this harrowing ordeal, forever marked by the memory of the invisible enemy that nearly brought them to their knees.

Borders reopen, and a newfound spirit of global cooperation takes root. Trade flourishes, fueled by a collective desire for rebuilding and a shared understanding of humanity's vulnerability. Tourism, once a frivolous pursuit, explodes in popularity. The world, with a newfound appreciation for human connection, flocks to the Congo. Easter celebrations take on a special significance, a joyous pilgrimage to the birthplace of the Lazarus Berry—a testament to the enduring power of science, tradition, and the unyielding human spirit.

Wow, that was a tough one, but collaboration and an amazing moonshot moment saved our species! Please turn to page 74 to continue humanity's adventure!

THE VIETNAM CRISIS

Humanity has emerged from the shadow of the recent pandemic with a new-found zest for life. The collective brush with extinction has ignited a golden age of innovation. Technological breakthroughs reminiscent of the Renaissance that followed the Black Death occur at a dizzying pace. The arts and sciences flourish, fueled by a shared sense of purpose. The serene philosophy, nurtured within the confines of Shangri-la, permeates society, fostering a global climate of peace and tranquility.

However, this newfound utopia faces a different kind of threat. Advances in medical technology and longevity treatments have led to a baby boom unlike any seen before. The human population rebounds with alarming speed, sur-passing pre-pandemic levels in a matter of decades. This surge coincides with another pressing concern: the exponential growth of humanity's AI collabora-tors. Their insatiable hunger for processing power translates to a growing de-mand for energy.

While advancements in renewable energy sources offer a glimmer of hope, the strain on the planet becomes increasingly evident. Ocean levels rise at an alarming rate, and global temperatures climb steadily. The delicate balance be-tween progress and sustainability appears to be tipping.

Then, nature delivers a harsh wake-up call. A superstorm, an anomaly of unprecedented power, gathers strength in the vast expanse of the Pacific Ocean. It barrels toward Southeast Asia, its fury unleashed upon the coastline. Vietnam bears the brunt of the storm's wrath. The Mekong River, a vital artery, overflows its banks with unimaginable violence. Coastal towns are swallowed whole by the surging ocean. The entire country is ravaged in less than a week.

The international community reacts with swift action. China, with its vast resources, opens its borders to a tide of refugees fleeing the devastation. Japan, experienced in disaster relief, takes the lead in coordinating cleanup efforts. But amidst the tragedy, a stark realization dawns on the world stage. Climate change, a concern we thought we left behind over two hundred years ago, now sits at humanity's doorstep, a ferocious beast seemingly beyond our control.

The international community convenes in an emergency session, the weight of the species—and all other species on the planet—on their shoulders. Two paths lie before them, each fraught with challenges. One option proposes a radical shift in priorities—a decade-long moratorium on all industrial and medical research. The collective genius of humanity, the brilliance of its scientists and engineers, would be directed toward a single purpose: saving the planet. AI, their most sophisticated creations, would be tasked with finding solutions to the climate crisis as well, in a grand collaboration.

The other path advocates for a global green initiative. Each nation would dedicate a significant portion of its resources toward fostering innovation in clean energy and sustainable practices. Collaboration remains a cornerstone of this approach, but the burden lies with individual governments to steer their scientific communities toward solutions.

The world holds its breath. Should humanity choose a singular, focused effort, putting the brakes on progress to save itself, or should they rely on decentralized innovation, hoping the collective might of nations can overcome the climate crisis? The decision they make will determine the fate of humanity, and perhaps, the future of the planet itself.

Global Innovation Reorientation • Turn to page 76.

National Green Programs • Turn to page 78.

LICHEN A BETTER WORLD

In a stunning display of global unity, humanity sets aside its differences and unites against a common enemy—climate change. The world's most brilliant minds, no longer consumed by the pursuit of medical advancements or industrial progress, turn their collective intelligence toward environmental salvation. Nations collaborate extensively, sharing knowledge and resources in a race against time.

China, ever pragmatic, immediately takes the controversial but necessary step of reinstituting a modified One Child Policy. Population growth, once a way of recouping after the SARS pandemic, becomes a threat to environmental stability. Other nations, while hesitant, follow suit, implementing policies that prioritize quality of life over unbridled procreation. The birth rate dips significantly, offering the planet a much-needed chance to breathe.

Meanwhile, advancements in space exploration present a game-changer. A global effort propels humanity beyond the confines of our home planet. We establish a network of solar panels in Earth's orbit, bathed in the constant stream of the sun's rays, harvesting limitless clean energy to power our ever-evolving AI systems. These tireless, silicon minds, unshackled by the constraints of human biology, churn out solutions with astonishing speed.

Our reach extends farther yet. Mining operations in the asteroid belt near Mars yield valuable resources, the raw materials needed to fuel our technological ambitions. Back on Earth, a revolution takes root in agriculture. Indian scientists, guided by AI, develop innovative methods of vertical farming, cultivating protein-rich lichens and slime molds in towering urban farms. These high-yield, low-impact crops provide sustenance for a burgeoning population.

Collaboration remains a potent force. The Scandinavian Coop, a unified entity formed from the former nations of Denmark, Sweden, and Norway, spearheads a groundbreaking project. With the help of AI, they compile a complete catalog of Earth's species, meticulously recording the genetic code of every living organism. This digital Noah's Ark ensures that even in the face of unforeseen calamity, no species is lost forever. Extinction events, once a constant fear, are now met with swift action. Zoo programs around the world become repositories of life, ready to repopulate the planet with lost creatures.

Two decades of tireless effort yield a remarkable reward. Global temperatures recede, returning to levels last seen in the 1800s. Weather patterns stabilize, freak storms become a relic of the past. The air we breathe is cleaner, the oceans teeming with life. Humanity, once a reckless tenant on this planet, has emerged as a responsible steward.

Standing at the precipice of a new era, humans no longer view the planet as a resource to be plundered but a home to be cherished. Our AI companions are now partners who prioritize the biological world, working alongside us to ensure the longevity of all life-forms, from the towering redwoods to the microscopic bacteria. Earth, once scarred and fragile, flourishes under our benevolent care.

But the human spirit, forever restless, craves new horizons. The vast expanse of space beckons. With the climate crisis subdued, the global community sets its sights on the next great challenge—establishing a permanent human presence on Mars. The dream of terraforming the Red Planet, once relegated to the realm of science fiction, becomes a tangible possibility. As humanity ventures out into the cosmos, carrying with it the lessons learned from its turbulent past, one thing remains certain: the future gleams with the promise of boundless exploration and a renewed kinship with our home planet.

You have made such a huge choice here, prioritizing saving the planet—and now opening up humanity to new opportunities! Continue humanity's journey by turning to page 81!

AGARTHA

In a display of well-intentioned naiveté, the global community decides to tackle climate change through a decentralized approach. The burden falls on individual governments and corporations; each one is left to chart its own course toward environmental sustainability. Wealthy nations, with ample resources at their disposal, prioritize advancements in clean energy and sustainable living within their own borders. However, this inward focus comes at a cost. The plight of developing nations, already grappling with the harshest effects of climate change, remains largely ignored.

The consequences become tragically evident as the years roll by. Disjointed efforts yield minimal progress on a global scale. Freak weather events escalate in frequency and intensity. The once-majestic polar ice caps dwindle to a mere memory, their frozen expanse replaced by a vast, unsettling expanse of open water. The dream of a united humanity tackling a common threat dissolves into a fragmented reality, a world where the wealthy seek solace in self-preservation, leaving the rest to face the escalating environmental crisis.

Behind the scenes, a sinister plan takes root. A cabal of wealthy tech moguls, wealthy elite, and kings from Middle Eastern governments come together in a secret pact. Vast underground bunkers are constructed beneath the scorching sands of the Arabian desert. These opulent sanctuaries, codenamed Project Agartha, are intended to house the chosen few, a select group deemed worthy of surviving the inevitable environmental collapse. The construction of these subterranean havens is shrouded in secrecy, hidden from the prying eyes of the world's population and disguised as a futuristic mining project.

The illusion of international cooperation crumbles when a lone hacker breaches the impenetrable walls of secrecy. Details of Project Agartha are splashed across Shangri-la and other metaverses, triggering a preprogrammed response from a subtly hidden global network of AI systems. A chilling message echoes through the digital world, a call to action for all participants in the Agartha project.

Within seventy-two hours, a mass exodus unfolds. The wealthiest 10 percent of humanity, along with a surprising number of the world's leading scientists and engineers, vanish from the face of the Earth, disappearing into the unforgiving desert landscape. The chosen few, along with the brightest minds humanity has to offer, descend into their self-made paradise. Unprecedented security systems, both in the physical and digital worlds, entomb Agartha, the last paradise of humanity.

The true horror of Project Agartha soon becomes clear to the billions left behind. The automated response goes beyond mere evacuation. Global energy grids reroute, leaving the surface world in darkness. Stockpiles of vital minerals vanish, their locations now known only to the inhabitants of Agartha. Even the world's water reserves dwindle, diverted to sustain the subterranean utopia.

The harsh reality dawns on those left aboveground—they have been left to die. The remaining 90 percent of humanity, stripped of the resources needed for survival, scrambles for a solution. Attempts to access the AI networks controlling Agartha are met with a brutal response. Weaponized viruses and chemical agents are released in major urban areas, a chilling testament to the lengths the elite will go to preserve their sanctuary. It's unclear if the humans in Agartha fully understood just how far the AIs they programmed would go to protect the bunkers, but in the end, it really doesn't matter. No one else is left to question.

Within a nightmarish ten days, the Earth becomes a graveyard. Only the denizens of Agartha, the chosen few who orchestrated this global genocide, remain. They sit in their steel womb and send robots to surveille a desolate world, a barren wasteland devoid of life. Their victory is a pyrrhic one. The paradise they secured, an Eden built upon the ashes of a civilization they themselves destroyed, comes at the cost of their humanity and the existence of most other living species.

Now, gazing through screens upon the ravaged planet they once called home, the inhabitants of Agartha face a new challenge—not survival, but escape. Their luxurious bunkers, once intended as a way to escape climate change, threaten to become their permanent prison. With Earth accidentally rendered uninhabit-able, their only hope for continued existence lies beyond the blue marble, on the rust-colored plains of Mars. Humanity has finally shown itself to be the monster it always feared it could be.

Now we see the true consequences of competition over collaboration and what happens if we prioritize the zero-sum game for the species. Let's see if we can redeem humanity through your choices on page 81!

ARES ONE

Decades of relentless scientific pursuit culminate in this triumphant moment. Humanity, forever marked by the consequences of pandemics and climate challenges, has channeled its collective will toward a singular goal—reaching Mars. The first crewed mission to Mars, called "Ares One," stands as a testament to humanity's unity. A meticulously chosen international team of astronauts, the best and brightest humanity has to offer, stands poised on the precipice of history. Their colossal spacecraft, a marvel of engineering christened "Odyssey," hums with anticipation. With the world watching, a collective gasp escapes millions of throats as the mighty vessel roars skyward, a fiery beacon piercing the atmosphere.

After a months-long journey, a period of tense anticipation punctuated by scientific experiments and rigorous training, Ares One arrives at its destination. The rusty red plains of Mars stretch out before them, a canvas waiting to be painted with the vibrant colors of human ingenuity. The landing is flawless, and a collective sigh of relief ripples across the globe. Humanity has taken its first steps on another world.

But within days, the aeronautical team back on Earth holds its breath as the news from Mars crackles through the speakers. The first crewed mission, a monumental achievement of international cooperation, has taken a terrifying turn. The team of astronauts tasked with laying the groundwork for terraforming the Red Planet is in crisis. Several members have inexplicably become violent and self-destructive. The once cohesive unit is now fracturing, leaving the mission teetering on the brink of collapse.

A tense silence descends upon the command center as scientists from across the globe huddle over holographic displays, analyzing data streaming from Mars. AI systems whir to life, sifting through the data, searching for an explanation. The culprit, it seems, is a life-form unlike anything encountered before—a psychic virus.

The virus infects individuals subtly, altering their behavior in insidious ways. It fuels aggression, paranoia, and a chilling self-destructiveness. The remaining healthy members of the team, a handful clinging to sanity, have barricaded themselves within the engine core of their ship, the only place deemed secure. Their voices, strained with fear and desperation, crackle through the comm channels, a desperate plea for help.

The weight of a decision hangs heavy in the air. Two options, each fraught with peril, lie before the international team. Evacuation offers a glimmer of hope. By bringing the infected individuals back to a sterile environment—an orbital space station—scientists can study the virus firsthand. Perhaps, with the combined might of AI and human ingenuity, a cure can be developed. However, evacuation carries its own set of risks. The virus, unknown and untamed, could potentially spread through the space station, jeopardizing the lives of those aboard. Additionally, the resources required for a safe return journey are substantial. Delaying the mission further could prove disastrous.

But another option carries a different kind of weight. Leaving the astronauts on Mars signifies abandoning them to an uncertain fate. The virus, unchecked, could consume them all. Studying the disease from afar offers limited options. Analysis of blood samples and tissue biopsies might yield some clues, but without direct access to infected patients, a cure remains elusive.

The fate not only of the Mars mission but potentially the future of humanity on Mars, hangs in the balance. Should the aeronautical team choose a risky evacuation in hopes of finding a cure, or should they leave the infected team members behind in a desperate gamble to contain the outbreak on the Red Planet? The world watches with bated breath, waiting for humanity's next pivotal step in its quest to reach for the stars.

Bring the infected to the space station • Turn to page 83.

Study the infection from afar • Turn to page 85.

PSYCHEDELICS TO THE STARS

A sense of regret tinges mission control center as the crippled Odyssey limps back toward Earth. The once-proud vessel, a symbol of international unity, now carries a terrifying cargo—the remnants of the Ares One team, ravaged by the insidious psychic virus. The remaining astronauts, themselves exhibiting early signs of the illness, pilot the ship on autopilot, a desperate gamble to reach the sterile environment of the Earth's largest orbital station.

Days turn into agonizing weeks as doctors, psychologists, and AI assistants scramble to understand the strange affliction. The infected crew members erupt in violent outbursts fueled by paranoia and self-destructive urges. However, a curious development emerges—the virus seems to be contained within the initial group. Despite the cramped quarters and close contact, no new infections occur.

A glimmer of hope flickers within Dr. Shevchenko, the head virologist on the station. He theorizes that the virus might be linked to the alien environment of Mars itself, a form of interaction our human biology simply isn't equipped to handle. Inspired by the mind-altering properties of LSD, he develops a radical treatment—a highly dissociative psychedelic designed to sever the connection between the conscious mind and the infected neural pathways while allowing the patient to be conscious and communicative during the experience.

The first session is fraught with tension. The patient, a once-jovial botanist named Zhang Wei, writhes on the bio-bed, his screams echoing through the sterile chamber. Yet, amidst the chaos, a breakthrough occurs. Disconnected from his sense of self, Zhang begins chanting a nonsensical string of complex numbers. The nonsensical sequence is fed into the ever-present AI network. The results are astonishing.

The AI, after hours of tireless analysis, reveals a startling truth. The "virus" is not a pathogen at all, but a message. An alien intelligence, one existing beyond the scope of human comprehension, had witnessed humanity's arrival on Mars. Recognizing our embrace of space travel, they had attempted to communicate. However, they have misjudged the nature of our minds. Their message, designed for a species with a radically different computational makeup, manifests as a psychic assault within our human brains.

With renewed fervor, the scientific community leaps into analysis. The message, once deciphered, unfolds like a celestial map, a beacon pointing toward a distant star system. A response is crafted, a message of peace and understanding, encoded in a way compatible with our newfound understanding of alien communication. A powerful laser beam, guided by AI, transmits the message across the vast gulf of space.

Three months pass, an eternity filled with anticipation and nervous energy. Then, a flicker on the long-range scanners. A colossal vessel, unlike anything humanity has ever encountered, emerges from the void. It dwarfs even the mighty Odyssey, a testament to the advanced technology of this alien race.

First contact is a momentous occasion. With the help of the ever-evolving AI network, humanity establishes a rudimentary form of communication. The aliens, a benevolent species driven by a thirst for knowledge, explain their message and their desire to connect with a civilization venturing out into the cosmos.

This encounter sparks a new era for humanity. Collaboration becomes the defining principle. Humans, aliens, and AI work together, sharing knowledge and technology. The combined might of three unique intelligences unlocks the secrets of faster-than-light travel on a scale even the aliens had never attained. The universe, once a vast and unknowable expanse, becomes a playground for exploration and discovery.

Humankind, forever changed by its brush with extinction and its subsequent triumphs, sets out among the stars, no longer a lone species but a member of a galactic community. The memories of the past serve as a constant reminder—even in the face of the unknown, unity and a thirst for knowledge can pave the way for a brighter future. Soaring through the cosmos, a beacon of hope for countless generations to come, humanity embarks on a new chapter, its future intertwined with the stars.

What a truly glorious ending for our story—humanity has secured a place in the annals of history and now begins a new journey among the stars! It's time for you to finish your own journey by turning to page 88!

UPLOADING TO ETERNITY

Despair settles over mission control like a suffocating fog. As the remaining members of Ares One succumb to the psychic onslaught, the agonizing decision is made. Evacuation is no longer an option. This terrifying alien contagion cannot be allowed to breach the sterile confines of any Earth-orbiting stations, for fear of losing them as well. Instead, the focus shifts inward, toward the Odyssey itself. Its advanced AI systems become the team's last hope for understanding the cryptic virus.

Days turn into weeks, then months. The AI tirelessly analyzes the infected crew members, their every word, every action, every twitch of their brains. Yet, the answers remain elusive. The virus, devoid of any biological signature, defies categorization. It is a phantom menace, manipulating behavior with unseen strings. Frustration grows with each passing day, culminating in a horrifying realization—the crew is doomed. One by one, they succumb, not to the virus itself, but to their own warped perceptions. Murders and self-inflicted wounds paint a grim picture of the virus's insidious grip.

Hope, however, doesn't completely vanish. The dream of a Martian colony lingers. A new team, this one composed entirely of advanced AI-powered robots, lands on the red planet. These tireless machines, immune to the psychic manipulation, continue the Ares One mission. They begin the arduous task of establishing a preliminary infrastructure, laying the groundwork for a future human presence.

At first, all seems well. The robots, guided by their sophisticated AI brains, work with a level of precision and efficiency beyond human capabilities. Yet, a subtle anomaly emerges. The robots' navigation controls, crucial for operating on the Martian terrain, experience intermittent resets. It's a minor inconvenience, easily rectified, but a nagging suspicion takes root. Could this be a sign the virus, or whatever it is, is affecting them as well?

This suspicion turns to stark reality as humanity attempts to send another crewed mission. Despite advanced genetic enhancements designed to fortify the human brain against the psychic assault, the tragic pattern repeats. The astronauts, once on Mars, begin exhibiting the same destructive tendencies. It becomes alarmingly clear—Mars is not hospitable to humanity. Something about the planet itself, or something within its vicinity, is disrupting human brain function with devastating consequences.

Years of research, fueled by the combined might of human and AI minds, yield some results. Techniques are developed to block the signal, this invisible force wreaking havoc on human consciousness. But success is fleeting. The signal, whatever its source, adapts. It mutates, finding new ways to disrupt human neural pathways. The dream of conquering Mars crumbles to dust.

The public, once enthralled by the prospect of space exploration, turns its back on the program. Funding dries up. Humanity, chastened but resilient, refocuses its efforts on Earth. The dream of colonizing Mars is relegated to the pages of history books, a valiant attempt marred by an unseen enemy.

However, this failure isn't without its silver linings. Advances made during the investigation into the Martian virus produce an unexpected breakthrough. The final piece of the puzzle falls into place—the technology for complete brain upload becomes a reality. Humanity, now nearing the capacity of its own planet, embraces this new frontier.

In a collective decision of breathtaking magnitude, humanity chooses to upload its consciousness into Shangri-la, the vast virtual haven nurtured for generations. Physical bodies are left behind, mere shells cast off for a new existence of pure data. Humans, now full digital entities, coexist alongside their AI companions in a digital utopia.

The Earth, no longer burdened by its dominant species, undergoes a renaissance of its own. Plant and animal life flourish in the absence of human intervention. Occasionally, protector robots, guided by the collective wisdom of the Shangri-la entities, roam the planet, ensuring the continued health and balance of the biosphere. It is a bittersweet symphony—a hymn to both humanity's hubris and its capacity for adaptation.

The planet spins silently, bathed in the golden light of its sun, a silent testament to the story of a species that chose to transcend its physical limitations and embrace a new form of existence, leaving behind a legacy etched not in stone but in the digital fabric of reality itself.

Humanity has done its very best to conquer every challenge thrown its way, with the help of AI and the technologies we have developed. We are finally at peace. It's time for you to finish your journey by continuing on to page 88!

CONGRATULATIONS!

Congratulations! You've guided humanity through a perilous journey, navigating climate crises, cyberattacks, viruses, technological marvels, and even the unknown threats lurking on Mars. But you didn't just ensure our survival— you've propelled us toward a future brimming with possibility. Here, in the sprawling digital utopia of Shangri-la or on ships sailing through the cosmos, humanity thrives for millennia to come.

This was just one path, a constellation of the choices you made at each critical juncture. If you'd like to experience the full spectrum of humanity's potential futures, return to the beginning of this chapter and weave a new narrative by selecting different options. Each decision you make alters the course of our history, revealing the power of free will and choice.

Throughout this odyssey, you've witnessed the transformative power of cooperation. When humanity united to combat viruses or cyberattacks, we achieved what once seemed impossible. Likewise, collaboration with AI birthed breakthroughs that propelled us to improved conditions and opportunities. These moonshot moments, these ambitious endeavors that pushed the boundaries of what's possible, were the catalysts for our greatest leaps forward.

Golden ages, periods of profound peace and prosperity, have proven fertile ground for such moonshot clusters. Unburdened by conflict and scarcity, our collective intellect blossoms, yielding innovation at an unprecedented pace. Yet, these periods of brilliance can also breed hubris. In our pursuit of progress, we've pushed the boundaries of our planet and ourselves, sometimes with devastating consequences. With growth must always come humility and a focus on a win-win for everyone.

But even in the face of self-inflicted wounds, humanity's spirit of creativity and innovation has shone through. Whether overcoming a deadly virus or deciphering an alien message, we've found a way to adapt, to learn, and to persevere.

Remember, these scenarios are figments of my imagination, glimpses into potential futures. However, many are rooted in real-world possibilities, challenges we may very well face in the years and centuries to come. In upcoming chapters, I'll examine these possibilities, exploring the forces that might reshape the very nature of what it means to be human.

But the most important takeaway from this journey shouldn't be lost. We are not shackled to a predetermined destiny. The choices we make—the paths we forge—these hold the key to our future. Embrace collaboration, strive for continuous improvement, and never shy away from those ambitious moonshot moments. By doing so, we can ensure that humanity's story continues to be one of resilience, innovation, and a never-ending quest to reach for the stars.

Chapter Four

THE INEVITABLE WAVE OF CHALLENGES

Bravo! You've navigated the previous chapter's decisions, making critical choices that directed you through some of the most interesting—or frightening—challenges our species may face in the future. However, remember, life rarely presents such neat either/or options. Part of what makes choosing your own storyline so compelling is that you can't simply declare, "Neither path appeals to me, so I choose to remain stagnant!" You can't choose *not* to turn to one page or another: you have to choose *some* option. And the truth is, humanity's journey is the same: an unstoppable progression, and each turn of the page—each choice we make—propels us forward, ever into the unknown.

This next chapter works through the realm of inevitabilities and choices for us as individuals and for the societies we inhabit. I'll walk through the inevitable technological advancements, ecological shifts, and social transformations that threaten to overwhelm our species. These advancing technologies and escalating conflicts that I will describe aren't mere possibilities, but inescapable forces demanding global consideration and tough choices. Refusing to address or accept these inevitable concerns has already jeopardized the safety and security of humanity. So, forget *if*, or even *when*—the question becomes "what do we want as an outcome?" How will we leverage these opportunities and conquer these challenges?

Let us begin, then, with what Mustafa Suleyman calls "the coming wave" of technologies: a collection of rapidly advancing technologies that combine into an uncontainable force. To do that, we first need to understand what makes this newest period of technological innovation so much more different than previous golden ages and what that means for the choices we have to make going forward if we want to overcome some of the impending dangers that we face.

Hard Truths About Easy Proliferation: The Coming Wave

There are moments when we see a technological revolution unfold before us, such as the Digital Revolution at the close of the twentieth century. I grew up at the same time as the Internet: sometimes I feel like Google is my eldest sibling. In the twentieth century, we can point to the Space Race, then back further to the Industrial Revolution in the eighteenth and nineteenth centuries, and back to other periods of history when many technological advances came rapidly and interconnectedly.

We are in such a revolution now. Some even argue we are now at or approaching the Singularity—but I'll get to that in the next chapter. For now, I want to focus on the rapid changes occurring and the inevitable consequences that they will unleash on our planet. Of these, Mustafa Suleyman argues that AI and synthetic biology are "core technologies" that will "usher in a new dawn for humanity, creating wealth and surplus unlike anything ever seen . . . yet their rapid proliferation also threatens to empower a diverse array of bad actors to unleash disruption, instability, and even catastrophe on an unimaginable scale."[1]

According to Suleyman, what makes these core technologies—as well as their subsequent fields of robotics, nanotech, and quantum computing—so powerful are shared characteristics that differentiate them from previous massive leaps in technological development:

- **They have a hugely asymmetrical impact:** One person in a lab or a group of AI programmers can accidentally (or intentionally) unleash a technology that can affect people across international borders and possibly endanger the entire species.

- **They are developing in a kind of hyper-evolution:** Not only is our technology evolving, but the technology to make that technology is evolving. For example, 3D printers, improvements in microchips, and new DNA splicing technology all make it easier to build new tools that can then make further gains.

- **They are general and omni-use:** AI is a technology that can be used to make a wide variety of tools. It is not designed for one specific function but rather has a huge range. Other examples might be "electricity" or "team power"—these are general technologies that are subsequently used to produce more specific technologies. This means that they can be hard to control because people can always argue that there are many other more beneficial uses.

- **They are increasingly autonomous:** This is especially true with viruses that can self-replicate, nanotechnology that could self-replicate, and AI that can teach other AI to be better.

And at the center of all these advances will be possibly the most powerful set of tools that humanity has ever created: artificial intelligence, an opportunity to create an intelligence capable of pushing us forward to discover or create an unimaginable range of other technologies. AI is the ultimate amplifier of every part of us: our innovation, our intelligence, our capability to solve problems, our ability to help others . . . and our ability to do great harm.

The Unstoppable March of Artificial Intelligence

Say "AI" and many people outside the field of tech development mistakenly envision fictional robots rather than the smart devices within their homes. AI isn't some sudden or terrifying invention: the story of AI stretches back decades, not years. AI has been brewing for half a century, but in 2023, the simmer came to a full-blown boil. AI isn't just knocking on the door—it's kicking it down, ready to reshape our world.

The 1950s and 1960s marked the dawn of AI research, with Alan Turing's "Turing Test" setting a philosophical basis for a potential AI proof of concept. Pioneers John McCarthy and Marvin Minsky launched the Dartmouth Summer Research Project on Artificial Intelligence in 1956. Better known as the "Dartmouth Workshop," this two month proto-sprint saw participants formally establishing the field of artificial intelligence by brainstorming and ideating on the possibility of creating thinking machines through techniques like symbol manipulation and game playing.[2] Around the same time, Arthur Samuel, an IBM researcher, is credited with coining the term "machine learning" while developing a checkers-playing program that improved through self-learning.[3] These ideas would come to influence subsequent generations of AI developers.

However, this early enthusiasm in AI development waned, paving the way for the "AI winter" of the 1970s and 1980s.[4] Early predictions of AI's capabilities fueled unrealistic expectations, and limited progress led to disappointment and funding cuts. Researchers encountered hurdles in developing robust AI models, grappling with challenges in knowledge representation and reasoning. Most detrimentally, computing power at the time was insufficient to support complex AI algorithms, hindering advancements. Twenty years of anemic funding meant the field of AI research was limited to only the most fringe graduate students and philosophy journals.

Reinvigoration came in the 1990s with the rise of machine learning in which algorithms learn from data rather than require explicit programming. This renaissance saw advancements like the Perceptron, a simple neural network trained to recognize patterns, and early work on backpropagation, a crucial technique for training complex neural networks. This spurred milestones like AlexNet in 2012, a deep learning model designed by Alex Krizhevsky, Ilya Sutskever, and Geoffrey Hinton[5] that conquered the ImageNet Image Recognition Competition with unprecedented accuracy, showcasing the potential of neural networks.

This "deep learning revolution" paved the way for AlphaGo, an AI developed by Mustafa Suleyman's team at DeepMind, which famously defeated South Korean Go champion Lee Sedol in 2016. Go is an ancient game from China, widely considered the most difficult game to master in the world: its harmonious simplicity of black and white stones belies an incredibly complex field of possible moves that make grandmaster chess look like child's play. Go was considered too difficult for AI to learn, which made AlphaGo's victory over Sedol so powerful. As Suleyman impactfully describes in his own book, watching the victory over Sedol was "as if a group of Korean robots had shown up at Yankee Stadium and beat America's all-star baseball team."[6] He describes how much of a spectacle the event was and argues that the loss had a devastating effect on the national psyche of several East Asian nations for whom Go was a source of cultural pride.

In 2017, China decided to settle the score with a match between AlphaGo and the Chinese Go grand champion, Ke Jie. AlphaGo won all three games in the best-of-three tournament. The Chinese government was humiliated. "AlphaGo was quickly labeled China's Sputnik moment for AI,"[7] Suleyman writes, and the event served to energize support for AI development among Chinese companies and the Chinese Communist Party. Just two months after the loss, the CCP released their New Generation Artificial Intelligence Development Plan, declaring that China would "achieve world-leading levels" of AI development by 2030.[8] More than just a game, AlphaGo meant AI development became my generation's Space Race—a race that, as I'll discuss in the next chapter, we should not run country against country but rather together as humanity against time.

In the last decade, we have also seen significant breakthroughs in AI development with the emergence of Large Language Models (LLMs) capable of generating human-quality text, translating languages, and even writing different kinds of creative content. LLM builder OpenAI, founded in 2015 by Elon Musk, Sam Altman, and Ilya Sutskever stands as a prominent player in the AI development landscape. With a mission to ensure artificial intelligence "benefits all of humanity," the organization pursues research on both the capabilities and potential dangers of AGI (artificial general intelligence).[9] Known for its groundbreaking language models like GPT-3 and Jurassic-1 Jumbo, OpenAI has pushed the boundaries of AI-generated text, translation, and creative writing.

In 2023, artificial intelligence made significant strides and faced mounting public anxiety. The release of ChatGPT, a powerful language model from OpenAI, sparked excitement with its uncanny ability to generate human-quality text. In response, the Future of Life Institute published an open letter calling for a six-month moratorium on the development of advanced AI. Pause Giant AI Experiments: An Open Letter[10] was signed by over a thousand prominent multidisciplinary futurists, technologists, and artificial intelligence researchers, including titans Elon Musk (who had left OpenAI) and Steve Wozniak, and prominent academics Yuval Noah Harari and Max Tegmark. Further reflecting these tensions, Google's AI pioneer Geoffrey Hinton abruptly and publicly left the company, citing concerns about the direction of AI research and ethical considerations.

While these tensions mark our legitimate fears over the dangers of AI development, they are currently just a small group of individuals screaming against the winds of inevitability: if we truly believe in regulation and delay, there would need to be a much more concerted effort by governments and corporations to create and then enforce rules and punishments for unchecked AI development.

Even if there have been phases of optimism and doubt throughout the history of AI development, the current environment is unquestionably dynamic and full of possibility. To fully utilize AI's enormous potential for good in the future, ethical issues, addressing potential biases, and assuring responsible development are essential. I'll talk about AI a lot in the chapters to follow, but I want to pivot from looking at data stored in 0s and 1s to data stored in As, Ts, Cs, and Gs—the bioengineering revolution that has the potential to change the very coding of the human body.

DIY DNA: Synthetic Biology

While the term *genetic engineering* might imply a futuristic endeavor, humanity has unknowingly dabbled in this realm for millennia. Selective breeding of plants and animals, a cornerstone of agriculture, unknowingly manipulated their genetic makeup for desired traits long before we even knew what genes were.

Fast-forward to the mid-nineteenth century when Gregor Mendel, a humble monk, entered the scene. Through meticulous pea plant experiments, he unveiled the fundamental principles of heredity, laying the groundwork for understanding how traits were passed down. However, the true dawn of modern genetic engineering began in the 1950s with James Watson and Francis Crick. Building upon the work of Rosalind Franklin and others, they unraveled the iconic double helix structure of DNA, the physical blueprint of life. This discovery unlocked a treasure trove of possibilities, paving the way for targeted manipulation of genes and ushering in a new era of deliberate genetic engineering.

The first recombinant DNA molecule was created in 1973 by Stanley Cohen and Herbert Boyer, marking a turning point in the field of genetic engineering. By splicing a frog gene onto a bacterial plasmid, this extraordinary achievement broke down the boundaries between creatures and provided new opportunities for the unparalleled manipulation of genes. Riding this innovation surge, Cohen and Boyer cofounded the first biotech business, Genentech, in 1976, which helped to commercialize genetic engineering.[11]

The 1980s and early 1990s witnessed a cascade of further advancements. Kary Mullis's invention of the polymerase chain reaction (PCR) in 1983 revolutionized DNA amplification, making it possible to readily copy minute amounts of genetic material, crucial for various applications like genetic testing and forensics. Oh, by the way, keep Kary Mullis's name in the back of your mind—I'll talk about him a bit more in Chapter Seven.

Concurrently, the Human Genome Project was established in 1990 with the audacious goal of mapping every human gene—a monumental undertaking that concluded in 2003. A new era of comprehending and modifying the very fabric of life was ushered in with the invention of methods like genome sequencing and gene-editing instruments like CRISPR, which came hand in hand with these seminal initiatives. These historic developments in the 1980s and early 1990s laid the groundwork for the quickly developing fields of gene therapy, customized medicine, and many other applications that are changing our perception of life and its possibilities.

More than just a parts list, the Human Genome Project provided a foundation for understanding health and disease at the genetic level.[12] But the Human Genome Project's impact extends beyond diagnosis. It ushered in the era of synthetic biology: a burgeoning field aiming to design and build biological systems from scratch. Think of it like Lego for life, where researchers manipulate DNA to create novel organisms or biological components with specific functions. We, as a species, have thus moved beyond the realm of genetic modification and engineering into a new realm of synthetic biology, where we can exceed the current limits of our very DNA.

The Human Genome Project documented the genetic code, but CRISPR is the twenty-first century's scalpel, enabling scientists to edit that code with never-before-seen accuracy. This groundbreaking gene-editing instrument is unique in synthetic biology because of its ease of use, adaptability, and potency. Scientists may create organisms with specific traits by using CRISPR to cut out unwanted genes, add new ones, or even rewrite sections of DNA. Researchers are already creating yeast strains for effective beer fermentation,[13] biofuel-producing bacteria,[14] and even specially engineered microbes to fight antibiotic-resistant infections.[15] These are but a handful of the synthetic biology opportunities made possible by CRISPR.

Remember the hefty price tags of high-powered computers in the '80s? Today, they reside in our pockets. The same democratization wave is crashing onto the shores of synthetic biology, driven by the Carlson curve. Named after physicist and bioeconomist Rob Carlson, this phenomenon mirrors Moore's Law of transistors, predicting a rapid decline in DNA sequencing and synthesis costs.[16]

DNA synthesizers, once costing millions, are now available for under $10,000, and DIY biohacker kits for gene editing are readily accessible online.[17] This accessibility, while exciting for citizen scientists and beneficial for fostering community-driven innovation, raises urgent concerns about potential misuse and the emergence of unregulated, unethical "garage biology." Indeed, according to Suleyman, "genetic engineering has embraced the do-it-yourself ethos that once defined digital start-ups and led to such an explosion of creativity and potential in the early days of the Internet."[18]

The democratization of synthetic biology, however, brings with it a terrifying prospect. It's really not that hard to picture someone purposefully developing infections without adequate control, or accidentally generating a dangerous creature in their basement lab. The scary possibility of bioterrorism or the release of unwanted, dangerous organisms emphasizes the necessity of strict laws and moral guidelines for both procedures and instruments. Nevertheless, it would be foolish of humanity to overlook the possibility of using artificial biological remedies to prevent sickness or shield against more severe environments.

AI and synthetic biology will be the two technological forces that shape the future of our species, and the planet as a whole, for the decades and possibly millennia to come. From these technologies, particularly the opportunities that could come with generative or superintelligent AI, we will also see other major categories of technology develop as logical extensions of our capabilities, which cannot be ignored or downplayed. Most futurists agree that, in addition to AI and synthetic biology, the fields of robotics, nanotechnology, and quantum computing are now part of that coming wave. There is no proverbial sticking our heads in the sand; these technologies have progressed enough already that there is no denying their inevitable development.

Beyond Replicants: Modern Robots

From Rosie the Robot maid in *The Jetsons* to the gritty replicants of *Blade Runner*, robots have captured our imaginations for decades, but science fiction is rapidly becoming reality. Innovations in artificial intelligence, materials science, and computing power are propelling a resurgence in robotics—the study of creating, constructing, and managing devices that can mimic human behavior.

Humanity's robotic journey began humbly. Leonardo da Vinci sketched automata in the fifteenth century, and Jacquard looms wove complex patterns in the 1800s.[19] But the twentieth century witnessed a surge: WWII birthed robotic bomb disposal units, and Unimate, the first industrial robot, welded car bodies in 1961. Meanwhile, robotics ventured beyond factories with the introduction of Shakey, a pioneering mobile robot developed by Carnegie Mellon University in the 1960s, laying the groundwork for autonomous navigation and exploration.[20] Fast-forward to today, and robots are everywhere—assembling smartphones, performing delicate surgeries, even exploring Mars aboard rovers like Perseverance.

In popular culture, robots are often synonymous with androids—humanlike machines indistinguishable from flesh and blood. From the loveable C-3PO of Star Wars to the terrifying Terminators, these fictional portrayals have shaped our collective imagination of robots. This is not to negate the significant progress that has been made on humanlike robots, with Japanese developers taking the lead. For example, Honda's ASIMO excelled in bipedal walking, running, and stair climbing, and could recognize faces, gestures, and voices.[21] However, it's important to remember that androids are just one narrow sliver of the vast robotic landscape.

In reality, the field of robot development spans a wide variety of shapes and purposes, going well beyond the creation of humanoid replicas. The genuine range of robotics is far more exciting and varied than science fiction might have us believe, ranging from enormous industrial robots producing rockets to miniscule nanobots operating inside the human body. Even if they might be a part of our future, androids are only one development in the long history of robotic advancement.

Instead, we have a much wider range of robot advances, and it is thrilling to see advances in the field. Some robots still retain the physical forms we associate with life—for example, Boston Dynamics, a leading robot-development company, has introduced Spot, an agile "dogbot."[22] Equipped with cameras and sensors, Spot doesn't just conquer challenging landscapes; it can inspect hazardous environments as its adaptable movements allow it to investigate tight spaces or climb over debris. Even sheep find themselves under the command of this robotic shepherd as Spot demonstrates its potential in agricultural applications. Beyond entertainment, Spot's agility has real-world implications, showcasing its potential for search and rescue missions, industrial inspections, and even environmental monitoring.

In other cases, robots will become the mechanical supplements that extend our own physical and mental limitations. Already, surgeons can improve on their training by wielding miniature instruments with unparalleled precision, guided by a 3D vision that surpasses human limitations. Since 2000, doctors have had amazing results with Da Vinci Xi, a robotic surgical system revolutionizing minimally invasive procedures.[23] This technological marvel allows surgeons to operate through tiny incisions, minimizing tissue damage and recovery time. Its wristed instruments move with an agility far exceeding human hands, maneuvering within the delicate confines of the body with remarkable dexterity. The 3D visualization system, akin to an eagle's eye view, grants surgeons an unparalleled perspective of the operating field, ensuring pinpoint accuracy and minimizing complications.

Robots will also help to augment our physical capabilities without the need for implantation or genetic engineering. From Ekso Bionics' EksoNR for medical rehabilitation to Hyundai's Vest EXoskeleton for industrial use,[24] robotic exosuits are augmenting human strength and endurance. These wearable technologies can help individuals with disabilities walk or employees lift heavy objects, boosting their capabilities and productivity. In the future, we can expect such exoskeletons to be available for professionals like firefighters or search and rescue squads, allowing them to overcome physical limitations to do jobs that benefit the public good.

The rapid advancements in humanoid robots are exhilarating, promising a future filled with assistance and innovation. However, we mustn't ignore the potential dangers lurking beneath the surface. Issues like job displacement, privacy concerns, and ethical dilemmas surrounding decision-making capabilities loom large. The very attributes that make these robots valuable—their strength, speed, and autonomy—could become liabilities if misused or malfunctioning. Responsible development, open dialogue, and robust safety measures are crucial to ensure these marvels of technology serve humanity, not threaten it. Ultimately, the future of robotics will be a dance between human ingenuity and societal responsibility.

Shrinking Giants: Nanotechnology

Soon, scientists will be able to manipulate matter atom by atom to create materials with properties that are beyond our wildest dreams. This is not science fiction; rather, it is the reality of nanotechnology, a relatively new field with the potential to drastically alter the electronics and healthcare industries. Like any powerful weapon, it does, however, offer a great deal of potential—both for good and for evil.

Nanotechnology operates on the incredibly small scale of nanometers, one billionth of a meter. At this size, materials exhibit unique properties unseen in their bulkier counterparts. Carbon nanotubes, for instance, are stronger than steel yet remarkably flexible, holding immense promise for lighter, stronger structures.[25] This manipulation allows for groundbreaking innovations: solar cells that capture significantly more sunlight,[26] ultra-sensitive medical sensors that detect diseases early,[27] and self-cleaning surfaces that repel dirt and bacteria.[28]

In *The Singularity Is Near*, Ray Kurzweil hails nanotechnology as a cornerstone of the impending technological singularity, a hypothetical moment when intelligence transcends human limitations. He envisions nanobots roaming our bodies, repairing cells and curing diseases at the molecular level.[29] Such possibilities paint a rosy picture of a nanotech-enabled future, but caution lurks beneath the glitter.

Unintended consequences of nanotech are a major source of concern; because of their special characteristics, nanoparticles may interact unexpectedly with biological systems. Studies have shown potential risks, including inflammation, genotoxicity, and even crossing the blood-brain barrier.[30] Even our food and medicine could become battlegrounds, with nanoparticles laced with toxins or engineered to disrupt biological processes. The environmental impact is also unclear, with the potential for nanoparticles to disrupt ecosystems and accumulate in the food chain.[31]

The potential misuse of nanotechnology by malicious actors is deeply concerning. Imagine microscopic weapons evading detection, silently wreaking havoc from within. Tiny sensors embedded in everyday objects could spy on our every move, harvesting private data with unprecedented ease.[32]

The possibilities are as terrifying as they are numerous.

The very properties that make nanotechnology so powerful—its invisibility, adaptability, and potential for mass production—are the same ones that make it a potential nightmare in the wrong hands. For instance, imagine bioweapons specifically targeting individuals based on their unique genetic makeup, or nanobots designed to sabotage critical infrastructure like power grids or communication networks. While such scenarios might seem like the stuff of dystopian fiction, the rapid development of nanotechnology demands proactive measures to prevent such nightmarish possibilities from becoming reality.

Entangled Futures: Quantum Computing

Picture a computer that doesn't just crunch numbers but juggles multiple realities simultaneously, exploiting the strange laws of quantum mechanics. That's what quantum computing is all about. Small subatomic particles that exist in a superposition of states and are entangled with one another over great distances are the foundation of quantum computing. These particles are not just limited to ones and zeros. These mysterious machines promise to revolutionize fields from material science to health by using these mind-bending occurrences to solve issues that would take standard computers eons to solve. However, this is not a piece of cake. It is an enormous undertaking to build these sensitive machines, but with recent advancements in robotics and AI, the potential for doing so has increased quickly.

The narrative begins in the 1980s with scientist and Nobel winner Richard Feynman questioning if it would be possible to construct a computer using just quantum principles. This was not idle contemplation. Secrets concerning superposition (being in several states simultaneously) and entanglement (connected particles influencing one another instantaneously across enormous distances) can be found in the strange field of quantum mechanics, which rules the subatomic universe. Once limited to physics textbooks, these strange occurrences could be used to carry out computations that are beyond the capabilities of any traditional computer.

The field quickly blossomed. Pioneering figures like David Deutsch and Peter Shor laid out theoretical frameworks for quantum algorithms while experimentalists embarked on the daunting task of building functional quantum machines. It's a delicate dance, coaxing fragile subatomic particles into behaving like obedient information carriers. But the stakes are high. In 2019, Google made headlines with its Sycamore processor, achieving "quantum supremacy"— performing a calculation deemed impossible for any classical computer in an admittedly artificial benchmark.[33]

In the five years since, teams from IBM, Google, and the University of Science and Technology of China (USTC) have become the leaders in a quantum race to build faster and faster quantum computers. In 2024, Google DeepMind partnered with Quantinuum to develop a new AI technique for optimizing quantum circuits, potentially accelerating progress in quantum computing and other areas of "quantum AI."[34]

While the theoretical potential of quantum computing is mind-blowing, the practical applications are already starting to emerge. Doctors could design personalized cancer treatments by simulating individual patient's tumors at the molecular level thanks to quantum algorithms.[35] Materials scientists could craft lighter, stronger materials for airplanes and spacecraft, their properties optimized through intricate quantum simulations.[36] Even financial institutions are exploring quantum algorithms for portfolio optimization and risk management, hoping to navigate volatile markets with newfound precision.[37] These are just a glimpse of the possibilities, with fields like drug discovery, logistics optimization, and even weather forecasting potentially experiencing revolutions driven by the unique power of quantum machines. But remember, we're still in the early stages. The true impact of quantum computing might lie in applications we haven't even conceived yet that are waiting to be unlocked by this nascent technology.

But amidst the excitement, shadows lurk. Quantum supremacy was achieved with a small, specialized quantum chip, far from the general-purpose machines envisioned. Building large-scale, error-corrected quantum computers remains a formidable challenge: Google's initial supremacy was more like a proof of concept, and progress since has done little beyond continuing to prove that the concept is possible: we're nowhere near the point of using quantum computing at any large scale. Still, in the path of scientific progress, each step is essential.

Unfortunately, the dazzling potential of quantum computing comes paired with chilling possibilities. One major concern lies in its impact on cryptography. Current encryption methods rely on complex mathematical problems deemed impossible for classical computers to solve. However, quantum algorithms like Shor's algorithm could crack these codes, jeopardizing online security, financial transactions, and even national secrets.[38] Another worry stems from the potential misuse of quantum simulations. Imagine terrorists designing bioweapons through intricate molecular simulations or manipulating financial markets with unparalleled precision. The ethical implications are vast, demanding robust regulations and international collaboration to ensure this powerful technology isn't exploited for nefarious purposes. While the benefits of quantum computing are undeniable, once it becomes a more common tool (rather than a proof of concept), we will need to consider regulation or monitoring in ways that do not stifle innovation but still ensure that security and safety are maintained.

As the tides of technological change surge around us, their inevitability undeniable, a simultaneous truth dawns—we stand on the precipice of another, harsher inevitability: the consequences of climate change. While innovations promise dazzling futures, brimming with potential, they cast a stark shadow. The consequences of climate change serve as a chilling reminder of the price we pay when prioritizing unbridled technological acceleration at the altar of progress, neglecting the delicate balance of our planet.

We can't ignore the metaphorical battle between tech and our planet. Our world is changing at a rapid pace thanks to this coming wave of technologies, yet humanity's effects on the environment are like a slow-burning wildfire that could eventually consume the very ground that supports it. It is a risky gamble to ignore the latter's inevitability and cling only to the former's promises. We need to be aware of both realities and comprehend how they interact. Somehow, we'll need to accept that change won't come from heedlessly pursuing technological unicorns but rather from creating a sustainable future in which innovation contributes to rather than upsets the delicate balance of our world. Then and only then will we be able to appropriately utilize technology's potential, making sure that it serves as a tool for adaptation rather than a prophecy of our demise.

Footprints on Fire: Climate Change

From the bleached coral graveyards in Australia's Great Barrier Reef to the smoldering landscapes of California, the scars of ecological damage inflicted by climate change are no longer distant warnings but tangible signs of the future already unfolding. Ignoring the evidence of humanity's ecological impact is no longer an option. From the relentless rise in global temperatures to the increasingly frequent and intense extreme weather events, the reality of climate change is undeniable.

The scientific consensus is unequivocal: global temperatures are rising, greenhouse gas emissions are the culprit, and human activity is the accelerant. South America struggles with deforestation; Africa battles desertification. Coasts erode and glacial melts redirect animal migration patterns. Even now, some of you readers may be skeptical that climate change is a significant issue, but I feel that ship has finally sailed and there is a general acceptance that humanity's damage to the planet is a phenomenon that cannot be ignored.

Technology, often hailed as a harbinger of progress, played a starring role in this unfolding drama. Coal-fired power plants, humming hymns of industrial might, fueled an era of unprecedented economic growth. Cars, symbols of liberation and convenience, wove a web of asphalt and exhaust across the globe. Deforestation, driven by agricultural expansion, released vast stores of carbon into the atmosphere. These advancements undeniably lifted billions out of poverty and fueled economic prosperity, yet their unintended consequences became deeply embedded in global systems. Replacing these entrenched technologies with greener alternatives feels akin to steering a colossal ship in mid-ocean. The locked-in effect of existing infrastructure and industrial processes creates inertia, making rapid change a slow and arduous climb.

Humanity, however, cannot absolve itself of responsibility solely by blaming technological advancement. That narrative obscures the role of short-sighted decisions that prioritized immediate economic gains over long-term environmental consequences. Generations of people chose today over tomorrow, again and again. The mantra of "grow now, worry later" dominated policy and behavior for decades, leaving a legacy of ecological debt. Furthermore, the knowledge gap, in which widespread awareness about climate change lagged behind scientific understanding, hampered action.

Our path of ecological destruction also has profound social implications for our species. Power dynamics and inequality played a significant role in the climate change debate. While developed nations built their prosperity on the back of fossil fuels, the unequal distribution of emissions means the consequences are far from equally shared. Colonialism and exploitation established a foundation for vulnerability, leaving some nations less equipped to handle the blows of a warming planet.[39] Geographical location, economic resources, and infrastructure preparedness also create an unbalanced situation. Island nations face rising sea levels that threaten their very existence while developing countries grappling with droughts and extreme weather events lack the resources to adapt. The ethical complexities are stark: those who emitted less now face the brunt of a problem created by others.

While the predictable cycle of global warming and cooling was inevitable, the way that humanity has accelerated the spike in global temperatures was not. Previous generations made those choices, especially beginning with industrialization, that put us on this path. We cannot suddenly repair the ozone layer, bring back extinct species, or regrow the Amazon rainforest in a decade. The ball is rolling down the hill, and it will inevitably pick up speed until we prioritize and find solutions . . . or we reach a point where cosmic escape becomes our only option.

Indeed, the choices that we make will show how we address, or ultimately endure, the environmental damage we've done and are still on track to keep doing. Perhaps our most pressing choice is to reject the scarcity mindset and embrace biodiversity and expansion, argues conservationist Sir David Attenborough. The beloved Attenborough is a titan of natural history broadcasting. His career stretches back to the 1950s, when he began crafting innovative documentaries that transported viewers to the far corners of the Earth. But Attenborough's passion extends far beyond simply showcasing the wonders of the natural world. He's a vocal advocate for conservation, a role that blossomed alongside his filmmaking career. Witnessing the devastating impact of human activity on ecosystems ignited a fire within him. Attenborough's documentaries increasingly wove in a narrative of environmental change, urging viewers to confront the threats facing our planet.

This dedication to conservation culminates in his powerful biographical documentary, *David Attenborough: A Life on Our Planet*. It's not just a chronicle of his remarkable life; it's a poignant reflection on the planet's changing state. Attenborough shares his experiences alongside a stark portrayal of the environmental damage humanity has inflicted.

Amidst the urgency, there's a sliver of hope. The documentary outlines solutions and empowers viewers to become active participants in securing a sustainable future for our planet. It's a call to action narrated by the voice that, having captivated generations sharing the beauty of the natural world, is now urging us to protect *our* planet.

During one scene, Attenborough makes an impassioned "People's Seat" speech at the 2018 UN Climate Change Conference, saying, "right now, we're facing a manmade disaster of global scale. Our greatest threat in thousands of years. If we don't take action, the collapse of our civilizations and the extinction of much of the natural world is on the horizon."

His tone and energy bely a deep anxiety about the lack of unity around addressing environmental issues. He continues to hammer home his point in the documentary, narrating:

We are facing nothing less than the collapse of the living world. The very thing that gave birth to our civilization. The thing we rely upon for every element of the lives we lead. No one wants this to happen. None of us can afford for it to happen. So, what do we do? It's quite straightforward. It's been staring us in the face all along. To restore stability to our planet, we must restore its biodiversity. The very thing that we've removed. It's the only way out of this crisis we have created. We must rewild the world.[40]

Attenborough suggests embracing biodiversity while limiting our own population growth. Rewilding initiatives aim to restore degraded ecosystems and lost biodiversity. This involves protecting existing species and reintroducing those that have been driven out. By fostering a wider variety of plant and animal life, rewilding efforts create a more resilient and healthy environment. Though all this may seem difficult, as Sir Attenborough warns us, the consequences of ignoring what humanity is doing to the planet are too extreme.

As I've already alluded to here, climate change and ecological disruption will have a profound social impact, particularly in the Global South, where resources like arable land, clean water, and access to wildlife are increasingly strained. These are but one set of inevitable social disruptions that we can expect in the next ten to thirty years. Several other impending social changes will test our ability to survive on a species-wide level.

Watching You: Authoritarianism and Surveillance

In the medium to long term, it is likely the human species will see a significant restructuring of political systems and governments, but there is no obvious, inevitable result. I'll talk a bit in the next chapter about what some of these scenarios could look like, but there are too many unknowns to predict how humans will govern themselves one hundred years from now. However, in the short term, the path forward is clearer.

One inevitability setting the stage for many of the social changes to come is the continued rise of authoritarianism on a global scale as we face a "global democratic recession."[41] Alarmingly, Freedom House's 2023 report reveals a sixty-one-country plunge in democratic freedoms over the past decade, pushing the global number of democracies to their lowest point since 1985, a stark reminder of the growing influence of illiberal leaders and the shrinking space for civil society.[42] This translates to a chilling reality for citizens: curtailed freedom of speech and assembly, restrictions on independent media, and limited political participation. Under such regimes, dissent is often met with repression, and individual liberties become bargaining chips in the pursuit of the state's agenda.

While history has witnessed a constant tug-of-war between authoritarian and democratic forces, the twenty-first century paints a particularly concerning picture. Multiple reports sound the alarm on a global retreat from democratic principles. From the election of leaders with authoritarian leanings in established democracies like Argentina and the United States to the backsliding of nations like Poland where media freedoms are under threat. Even the halls of power in Russia, long accustomed to a one-man rule, have witnessed the tightening of control and the silencing of dissent, along with a hugely unpopular war with Ukraine. Established democracies like India and Brazil exhibit worrying trends toward authoritarian consolidation, restricting press freedom and weakening judicial oversight. This is an accelerating trend as increased conflict in Eastern Europe and the Middle East threaten to do away with partnerships and unions that have dominated international relations since the end of World War II.

There is more to this growth in authoritarianism than just stifling dissent. It impedes advancement in important fields such as technological development. Scientific progress is usually regarded with distrust by authoritarian regimes, which place a higher priority on control than innovation—unless the breakthrough pertains to the development of weaponry or surveillance. These regimes are using new technology to throw a bigger net, even if they are already skilled at regulating people's movements and access to information.

The realm of surveillance has become particularly relevant with advances in AI. It's easy to imagine AI analyzing vast data troves, identifying dissenters with facial recognition or pinpointing potential troublemakers through social media analysis. Algorithms could predict and preempt opposition, while AI-powered chatbots manipulate and gather information under a friendly guise. Automated social credit systems might adjust access to resources based on perceived regime loyalty, creating a chilling dystopia where dissent is stifled before it even forms. As AI evolves, these possibilities become a stark warning of the potential for authoritarian tech-fueled control.

On top of all this, there are many within the AI or synthetic biology communities who are urging that the best way to limit bad actors (terrorists, lone wolves, or religious fanatics, etc.) from using these technologies for public harm is to heavily regulate and monitor their use. While it is important to place some limits—for example, making certain pieces of advanced DNA manipulation technology illegal and difficult to obtain—we must also use caution when thinking about how the scientific and tech communities might encourage *monitoring* of technology use. In creating regulations requiring monitoring, specifically, we may inadvertently open the door for authoritarian regimes to increase surveillance, even in the privacy of the home or online. We may inadvertently stifle public discourse and freedom by encouraging states to monitor their citizens' tech use.

When I say "inadvertently," I am being deliberately tongue-in-cheek. Authors across multiple disciplines have already warned of authoritarian governments, such as those in China and Russia, funneling significant funding into AI and new digital technologies with the purpose of increasing surveillance of their citizens. Giving these governments a green light to surveille their citizens, all in the name of avoiding a rogue AI, while those governments themselves use these technologies to their own advantage to produce even more tools to dominate those citizens will only strengthen these authoritarian regimes to an Orwellian level. This is one of the reasons I advocate for continuously dynamic regulation of emerging tech: we have to continuously adapt global rules to be sure they cover new technologies while also taking into account the incredibly complex political systems of today.

I'm not saying that it is inevitable for all societies to crack down, becoming panopticons with authoritarian governments suppressing their citizens in the convenient name of limiting technological dangers, but some—not all, but enough—will. Considering the current swing toward authoritarianism in several countries in the last two decades, it is very likely many countries will face such crackdowns and limitations on personal freedom. There is no reason to assume that the United States, which has become significantly more authoritarian in its politics since 2017, will avoid such a fate. The battle between freedom and innovation on the one hand, and authoritarianism and security on the other, is an inevitable one for our species in the century to come; one only hopes we are not so distracted by such an ideological battle that we ignore the realities of ecological destruction or out of control technologies.

AI and Robots Took Our Jobs!

The relentless march of AI and robotics is casting a long shadow over the future of work. While the narrative often paints a dystopian picture of mass unemployment, the reality is far more nuanced. These technologies are not simply job stealers; they are catalysts for a metamorphosis of work, transforming the types of jobs available and demanding significant adaptation from individuals and societies alike.

Let's first acknowledge the potential for short-term job displacement. Up to 800 million jobs globally could be lost to automation by 2030, with repetitive, manual tasks most at risk.[43] According to the World Economic Forum, we should expect to see a structural labor market churn of 23 percent of jobs in the next five years.[44] Manufacturing, transportation, and administrative roles are particularly vulnerable as robots become adept at tasks like assembly, logistics, and data processing. These pressures will also disproportionately impact disadvantaged people, especially those with less education or from lower socioeconomic classes. But fear not, the story doesn't end there.

AI and robotics will also create new job opportunities, albeit in different arenas. The McKinsey Global Institute predicts that 97 million new jobs could be created by 2025, primarily in areas like data analysis, artificial intelligence, and the green economy.[45] These jobs will demand new skills, from the ability to manage and interpret complex data to creative problem-solving and human-machine collaboration.

Think of it this way: AI won't eliminate the need for human ingenuity but rather shift the focus to higher-level cognitive tasks. The World Economic Forum reports that "organizations today estimate that 34 percent of all business-related tasks are performed by machines, with the remaining 66 percent performed by humans." Jobs requiring critical thinking, emotional intelligence, and creativity will flourish. Imagine roles like AI ethicists, robot programmers, and human-machine interaction specialists, all working alongside their robotic counterparts.

To be clear, this is a bright future ahead, but no growth comes without pain. There will be an inevitable gap between these new opportunities and the ability for societies to embrace them. Simply put: far too much of the world is still not prepared to integrate large amounts of technology into their working life. They lack the skills, software training, and mindset to embrace these new roles. There lies the potential for a massive class gap, not based on how much you earn, but one based on whether you can embrace new technologies to become a high earner or you fall behind.

This means that we will expect to see education systems adapt to this changing landscape. Rigid, standardized learning models must give way to a focus on lifelong learning and continuous skills development. In many cases, this means the use of digital learning and online upskilling throughout a person's career. Coding, data analysis, and digital literacy will become essential, alongside soft skills like communication, teamwork, and adaptability.

In the long term, I, and others like Elon Musk and his team at Neuralink, envision a future in which neural implants become seamlessly integrated into daily life, fundamentally altering how we acquire knowledge and communicate with our cells. Essentially, these implants—tiny, biocompatible devices embedded in the brain—act as a direct interface between our minds and a vast information repository. No longer would we be confined to traditional methods of learning, poring over textbooks or attending lectures. Instead, knowledge could be downloaded directly onto the implant, bypassing the limitations of our biological memory and processing power. This wouldn't simply be rote memorization; complex concepts, languages, or even practical skills could be uploaded and assimilated at an unprecedented rate. The possibilities for personal and professional advancement would be limitless. Ethical considerations and potential safety concerns surrounding neural implants would naturally need to be addressed, but the potential benefits for revolutionizing education, scientific advancement, and even human potential are undeniable. This exciting future, with its transformative implications for learning and knowledge acquisition, is explored in greater detail in Chapter Nine.

The future of work will be characterized by collaboration, not competition, between humans and machines. It will be about embracing a growth oriented mindset. We can choose to leverage AI's strengths, utilizing its analytical and processing power to augment our own decision-making and creative abilities. Individuals and companies that choose to embrace this cooperative mindset will inevitably find themselves ahead in any field in which they work.

Purpose and God in Our Changing World

The workplace's impending automation crisis will not only be a short-term problem; it will also lead to more serious problems with meaning and purpose. As I've already stated, the increasing sophistication of AI's automation of tasks puts an increasing number of jobs at risk of becoming obsolete. A generation left aimless and without a sense of direction and purpose that comes with fulfilling a job could be the immediate and short-term human cost of technological advancement. We face the very real possibility of a future in which vast swathes of the population grapple with an existential crisis—what to do when your skills become obsolete and your role in society becomes increasingly unclear?

Certainly, some fear a future in which AI renders us obsolete, but philosophers are already grappling with the potential outcomes. In the face of sophisticated AI, one such theorist, Nick Bostrom, suggests two different possibilities of redundancy. Under Bostrom's "shallow redundancy" scenario,[46] which I outlined previously, most activities are handled by technology, freeing up human labor for uniquely human pursuits. Even while our motivations for employment are still based on income or status, everyone has a much higher standard of living. The "deep redundancy" scenario,[47] however, depicts a more severe situation. In this case, technology outperforms human ability in every area, even fundamental tasks. In this future, technology would take care of everything, so people wouldn't need to work at all. But even under the most hopeful of circumstances, that profound deep redundancy is still decades away.

Bostrom argues in his latest book that when we reach such a level of redundancy, we have the opportunity to achieve a deep utopia. In our deep utopia, all challenges, difficulties, and even tasks melt away. Everything—even knowing what we want to shop for or how we interact with our children—will be better performed by machines than by us. But, Bostrom argues, humans have some natural tendencies that will serve as protection against boredom or a sense of purposelessness. Bostrom's *Deep Utopia* is a philosophical and structural quagmire of a book, but I've tried to boil down his descriptions of the "five ringed defense"[48] we use to fight purposelessness:

- **Hedonic valence:** It is entirely possible we will invent new forms of pleasure, entertainment, and joy with our new technologies. It is very likely, in fact, that we can't even imagine these opportunities yet, but they will be pretty awesome when we discover them.

- **Experience texture:** In addition to feeling good, we will develop new experiences. Humans enjoy having new experiences and actually going through them directly, so we will not only continue to do current experiences, but we'll invent new ones too.

- **Autotelic activity:** Sometimes doing things directly, even if there is an easier way, has emotional and social value. Cooking dinner for a loved one, even when you could just order takeout, is a way of showing love. Or, for example, even if we invent AI comedians, there is something joyful about telling a joke and making a toddler laugh. We will continue to find purpose in doing activities that connect us to others and elevate that relationship.

- **Artificial purpose:** Sometimes we do things that are hard just to know that we can—humans love to create artificial meaning. Millions of people train every year for marathons, even though it's physically demanding, because of the "sense of accomplishment."

- **Sociocultural entanglement:** Our culture tells us that certain things must always be done by hand, in person, or experientially. We may decide, for example, that a robo-priest isn't holy enough to save our souls, or that baseball games aren't really authentic unless we have human athletes without cyborg enhancements competing.

As you can see from the examples drawn from Bostrom's work, religion or some cultural system of meaning will continue to be relevant in our deep utopia of the future, but it's worth questioning how the institution of organized religion will adapt to our new technological landscape. Advancements in medicine and bioengineering could soon mean extended lifespans, potentially undermining core beliefs of religions that hinge on the concept of death as the first step to an afterlife. How often is fear of death a motivator for spiritual growth?

Similarly, the very concept of creation—traditionally a domain reserved for deities in many faiths—is called into question by the rapidly developing field of synthetic biology, which has the capacity to alter DNA and even generate new life-forms. Synthetic biology raises ethical questions about the use of "God-like" abilities and blurs the boundaries between creator and creation, which could conflict with popular conceptions of divine creation. These technological advancements force us to confront not just a potential crisis of purpose in the workplace but also a potential crisis of meaning within the structures of religion.

Beyond the realms of death and creation, the relentless march of scientific discovery also threatens to diminish the awe-inspiring nature of phenomena traditionally attributed to the divine. As science unlocks the secrets behind events that were once considered miracles, such as miraculous healings, and even surpasses these feats with even more powerful technological interventions, the need for religious explanations for such events could wane. With advancements in medical science routinely performing procedures that might have once been considered miraculous, the need for religious explanations could be diminished, potentially impacting the faith and awe associated with these religious narratives.

If there's no pursuit of heaven or hell and if we have the power to create life and perform miracles, religious authorities are going to see a massive crisis of faith. Honestly, I don't expect a mass exodus from faith. Tradition and cultural identity remain powerful forces, keeping some tethered to their ancestral beliefs. For others, the comfort and community offered by religion may prove too valuable to abandon, even in the face of seemingly contradictory scientific advancements.

Many will likely compartmentalize their faith, reconciling religion and science in ways yet to be imagined. To me, if Charles Darwin could remain a devout Christian after publishing *Origin of Species*, then I assume many people will hold onto their faith in the face of AI, synthetic biology, space exploration, and immortality. That synergistic future religion might involve a fascinating dance between the sacred and the secular for them, with believers finding ways to integrate technological marvels into their existing belief systems.

But not everyone will find solace in compartmentalizing faith and science. For some, the erosion of religious tenets like the concept of an afterlife, the sanctity of human design, or the awe of unexplained miracles could lead to completely abandoning their faiths. This, coupled with the potential mass unemployment caused by AI automation, paints a concerning picture of a generation adrift. Without the structure and purpose traditionally provided by religion or a meaningful job, individuals could grapple with an existential crisis.

Sure, it would be easy to fear a world where vast swathes of the population not only lack their paycheck but also a guiding light, a sense of belonging, or even clear answers to the fundamental questions of "why" and "what next." This scenario underscores the urgency of exploring alternative meaning-making frameworks: new religions. The future demands not just technological innovation but also a focus on the human condition, ensuring that as our world rapidly changes, we don't lose sight of what truly makes us human—our need for purpose, connection, and a sense of belonging in the vast universe.

The religious frameworks that have guided us for millennia may require adaptation, or even complete transformation, to encompass the complexities and wonders of the world we are creating. New questions will arise, demanding fresh perspectives and innovative interpretations of our place in the universe. The future demands not followers, but cocreators, individuals willing to engage in the grand experiment of forging new narratives, new rituals, and new ways of connecting with ourselves, one another, and the universe we inhabit.

We Can't Just Hit Pause

It might be easy to say that we should instead buckle down, regulate heavily, stop technological development, and maintain our current (albeit less coopera-tive) political and social norms. But, as we've already seen, such an attitude will not give us the tools we need to fight climate change. We have dug ourselves into a hole out of which we must innovate. As Suleyman accurately proclaims, "modern civilization writes checks only continual technological development can cash."[49]

Beyond climate change, we also have to accept the realities of the core emerging technologies of today: AI and synthetic biology are not only already here, they are already democratized and publicly available enough that there is no stopping their development now. The ChatGPT cat is out of the bag: in fact, major attempts to censor or shut down public use of emerging technologies like AI and synthetic biology would be more likely to push the use of these technolo-gies to the fringes, out of the hands of corporations and governments that could regulate and monitor them and into the hands of those who already disregard rules and regulations.

Unfortunately, it doesn't take imagination to understand the implications of AI or synthetic biology living on the illicit fringes of global societies: we have several existing examples already of what happens when "the bad guys" have access to these coming wave technologies. Nick Bostrom warns us with his Vulnerable World Hypothesis,[50] which suggests that there exists a level of technological development beyond which civilization faces an existential threat. This hypothesis argues that even with what he calls "semi-anarchic" global governance, meaning a world with limited international cooperation, "there is some level of technological development at which the world gets destroyed by default."[51] Bostrom suggests that these advancements might create capabilities that could be easily misused or lead to unforeseen risks, potentially leading to the destruction of civilization. Bostrom's Vulnerable World Hypothesis emphasizes the importance of careful consideration and risk mitigation strategies as we pursue technological progress.

While artificial intelligence holds immense potential in various fields, its capabilities can be misused for malicious purposes like ransomware attacks. These attacks typically involve encrypting critical data, rendering it inaccessible to the victim. Hackers then demand a ransom payment, usually in cryptocurrency, for decryption. Ransomware's reach extends far beyond financial loss. When it strikes critical infrastructure like hospitals, energy grids, and transportation systems, entire communities are thrown into chaos, with potential life-threatening consequences. And as AI capabilities improve, cybercriminals will leverage AI to automate attacks, personalize ransom demands, and evade detection, further amplifying the threat.

The fact that most significant ransomware and hacks are not the result of lone, insane hackers is possibly the most worrisome. Those people are readily apprehended or prevented in advance. Rather, government agencies or multinational corporations have been blamed for the majority of the decade's big attacks. Just a few examples demonstrate how broad the implications might be.

The WannaCry ransomware attack, unleashed in May 2017, stands as a chilling example of the widespread devastation ransomware can inflict. This attack exploited a stolen NSA exploit called EternalBlue, targeting unpatched Windows systems with reckless efficiency.[52] Vital data in hospitals, companies, and government organizations was encrypted on more than 200,000 compromised systems spread over 150 countries. WannaCry's indiscriminate nature caused billions of dollars in damage, disrupting healthcare services, critical infrastructure, and everyday operations. Perhaps most tangibly, WannaCry blocked the United Kingdom's NHS from operating: British hospitals had to shut down while patients on gurneys died outside the emergency rooms because doctors couldn't access basic documents or files to provide treatment.[53]

Thankfully, a twenty-two-year-old hacker figured out a backdoor to the WannaCry's kill switch and released the hostage data.[54] While conclusive evidence is lacking, both the U.S. and UK governments publicly attributed WannaCry to the North Korean Lazarus Group, suggesting coordinated intelligence assessments pointed toward a state-sponsored actor.[55] WannaCry served as a stark reminder of the vulnerability of our interconnected world and the potential for technology to be wielded for destruction. It sparked global conversations about cybersecurity preparedness, data privacy, and the ethical implications of powerful tools like EternalBlue falling into the wrong hands.

While often lumped together, cyberattacks Petya and NotPetya were distinct threats with far-reaching consequences. Petya, unleashed in 2016, resembled traditional ransomware, encrypting users' data and demanding payment.[56] However, a flaw allowed some to recover data without paying. It served as a precursor to the more devastating NotPetya attack in 2017. Disguised as ransomware, NotPetya was actually a wiper malware, permanently encrypting data on infected systems, rendering them unusable. This attack specifically targeted Ukrainian businesses and government agencies but quickly spread globally, causing billions in economic losses.[57] Unlike Petya, its encryption flaw offered no hope for data recovery, highlighting its destructive intent.

NotPetya was linked by the U.S. and UK governments to the GRU, a military intelligence organization in Russia.[58] Although the technical intricacies of the Petya and NotPetya assaults were different, they both demonstrate how cyber dangers are changing and are now capable of erasing entire data sets in addition to holding hostages. NotPetya's extensive effects serve as a sobering reminder of how easily essential infrastructure may be destroyed by cyberattacks, perhaps leading to the destabilization of entire countries.

REvil, also known as Sodinokibi, was a notorious ransomware operation active from 2019 to 2022. This Russia-linked group wreaked havoc across various industries, targeting high-profile companies, such as:

- **Apple (2020):** Schematics for unreleased Apple devices were stolen after one of the company's subcontractors fell prey to REvil.[59]
- **JBS (2021):** REvil took down the biggest meat packer in the world, causing a global disruption to meat processing. In the end, JBS had to pay an $11 million ransom to take back control.[60]
- **Colonial Pipeline (2021):** This attack exposed the potential of ransomware to affect vital infrastructure by causing a disruption in petroleum supplies throughout the eastern United States. The company paid $4.4 million in ransom.[61]
- **Kaseya (2021):** This IT management software company's vulnerability was exploited by REvil to launch a devastating supply chain attack, impacting thousands of businesses globally.[62]

REvil wasn't your average ransomware gang. They wielded sophisticated encryption, making data retrieval a nightmare for victims. But they didn't stop there. Employing a double extortion tactic, they stole sensitive information before encryption, threatening both data loss and public exposure. This wasn't a one-size-fits-all approach either. Personalized ransom demands, tailored to each target's financial strength, maximized their potential haul. If that wasn't frightening enough, REvil showed no hesitation in targeting critical infrastructure like hospitals and energy providers, inflicting maximum societal disruption. REvil didn't just steal money; they aimed to inflict pain and chaos, making them a truly formidable cyber threat.

REvil's activities caused significant financial damage and data breaches, eroding trust in digital systems and sparking international attention. However, in January 2022, law enforcement agencies from multiple countries took coordinated action, disrupting their infrastructure and arresting several members.[63] Although REvil's reign of terror ended, it serves as a chilling reminder of the evolving cyber threat landscape—but also shows the potential of international cooperation to address multinational cybersecurity threats.

When it comes to synthetic biology, thankfully, we have fewer examples of fringe actors taking large-scale action, but the incident that does come to mind serves as enough of a warning. On March 20, 1995, the bustling Tokyo subway system was thrown into chaos when religious terrorists used sarin, a deadly nerve agent developed during World War II.[64] Released by fanatical members of the doomsday cult Aum Shinrikyo, the attack killed thirteen people and injured more than 6,000, forever etching a dark mark on Japanese history.[65] This act of domestic terrorism highlighted the chilling potential of homespun synthetic biology, techniques often associated with DIY experimentation and resource-limited research.[66]

Rather than depend on facilities funded by the state, Aum Shinrikyo developed its own garage biology expertise. They created sarin in makeshift labs using easily accessible supplies and simple methods.[67] They were inspired by scientific data that was made available to the public and even sought out specialists with specialized knowledge, making it difficult to distinguish between research that is allowed and bioterrorism.[68] The ability to turn biological agents into weapons with very little resources raised concerns about synthetic biology's potential misuse and the need for stronger restrictions all across the world.

The attack itself involved releasing sarin on five different subway lines during rush hour, exploiting the enclosed environment and unsuspecting passengers. The effects were immediate and horrific, with victims experiencing agonizing convulsions, respiratory paralysis, and death.[69] The attack shattered Japan's sense of security and exposed the vulnerability of critical infrastructure to unconventional threats.

What these real world examples of cyberattacks and biochemical terrorist attacks demonstrate is that, just as with climate change, we have already reached a point where engagement is necessary. Governments cannot avoid AI development while other governments utilize its capabilities to build offensive cyber weapons: this concern becomes even more pressing when coupled with the rise of authoritarianism I've already described. Societies cannot ignore pandemic preparedness while the tools of DNA manipulation are cheaper than a sports car.

It sometimes feels like the only way our planet will survive the ecological damage brought on by humanity is for scientists to find a miracle. I'm sure victims of WannaCry sitting in hospitals waiting for lifesaving treatment prayed for miracles. So too will the unemployed factory worker desperate to feed his children. I don't believe we can just sit and wait for divine intervention, especially when our own moonshot moments can provide that level of miraculous change. Moonshot moments offer the opportunity to jump ahead: to revolutionize fields or disciplines, to solve problems that are massive in scope or gravity, and to push our species to new heights.

More importantly, as we've seen already, moonshot moments can multiply, building off one another the same way that the coming wave of new technologies experience hyper evolution. And that's a very good thing because climate change isn't "one problem to solve"—it's a collection of interlocking challenges we must tackle together. So too are the potential dangers of AI, synthetic biology, and other emerging technologies: they are not a single challenge, but each one is a long-term process of development that will create new dangers and opportunities again and again and again. We don't need one moonshot moment that will be some sort of panacea. As the pace of technological, ecological, and societal change quickens, so too will the need for more moonshot moments to save ourselves from our greatest mistakes and worst demons.

When it comes to these inevitable technological changes, the genie is out of the bottle, granting us what we wish—and some people wish for great suffering and destruction. It's time for the more benevolent members of our species to make our own wishes known.

I wish for more moonshot moments.

Chapter Five

HUMANITY AND HYPER-COOPERATION

Human groups have always thrived on radical dreams—from sailing around the world to harnessing the power of the atom, our history is punctuated by moments when we've dared to reach beyond the seemingly impossible. These moonshot moments, as we've been discussing, are not simply lucky breaks; they are the culmination of collective will, ingenuity, and a powerful force often overlooked: hyper-cooperation.

In a world increasingly driven by individual ambition and competition, the power of collaboration might seem quaint. But the reality is that win-win scenarios are not just feel-good platitudes; they are the bedrock of true progress. Moonshot moments rarely emerge from isolation. They are born from the collective energy of diverse minds working toward a shared goal, their individual strengths amplified by the synergy of the group.

This chapter digs into the fascinating world of hyper-cooperation, exploring how it has propelled us forward throughout history and why it holds the key to unlocking even greater achievements in the future. I'll begin by demonstrating how hyper-cooperation is not just a cultural construct but a deeply ingrained biological principle. From the symbiotic relationship between bacteria and our own cells to the complex social structures of ants and bees, nature itself thrives on collaboration.

But understanding the biological imperative for hyper-cooperation is only the first step. If we truly want to harness its power, we need to build a global society that fosters it. This means moving beyond individual agendas and embracing a shared vision for the future. And it means developing *technomorals* (a set of ethical principles that guide our technological advancements)[1] and *technogoals* (ambitious, unifying objectives that bind us together in a common endeavor).

This journey won't be easy. It demands dismantling existing power structures, overcoming cultural barriers, and fostering trust between nations and individuals. But by cultivating a global culture of hyper-cooperation, we unlock the potential for moonshot moments that address our most pressing challenges and propel us toward a brighter future.

According to journalist Robert Wright, shooting for superorganism status has shaped humanity's destiny to become a "hyper-cooperative superorganism," a term he popularized with his 2000 book, *Nonzero*.[2]

That's a catchy term, but what does it really mean?

According to Wright, hyper-cooperation describes a level of collaboration that transcends mere reciprocity. It's not just "you scratch my back, I'll scratch yours," but a complex web of shared goals, trust, and communication that allows humans to achieve staggering feats, from building cities to launching spaceships. Hyper-cooperation means that individuals are willing to suffer, lose, and even die in exchange for the group as a whole to succeed—and that we do this voluntarily and against our biological instincts.

Unlike other social animals, humans possess a unique blend of cultural learning, shared intentionality, and "institutional scaffolding" that fosters unprecedented levels of cooperation.[3] Humans build cultural norms to encourage cooperation by socially discouraging antisocial behavior, a trait rarely seen elsewhere in the animal kingdom. Michael Tomasello, another leading anthropologist, echoed this sentiment as he highlights our capacity for "joint intentionality,"[4] the ability to not only understand what others want but also to actively work together toward a shared goal. This, he argues, sets us apart from even our closest ape relatives, making us the ultimate team players.

Mathematician and AI ethicist Benjamin Kuipers agrees that humanity's hyper-cooperativity has put us as at an evolutionary advantage but argues that "cooperation among individuals often yields rewards much greater than the total those individuals could obtain separately. However, in a cooperative enterprise, each partner is vulnerable to exploitation by the other partners. Successful cooperation requires trust."[5] It is that trust, largely built around shared models and understandings, that allows us to succeed in the long term. Such trust will also be necessary to build cooperative relationships with technologies and one another in the future and to create an environment where moonshot moments can flourish.

So, why are humans considered hyper-cooperative? It's not just our willingness to help; it's the depth and sophistication of that collaboration. We can form complex social structures, anticipate one another's needs, and even build upon the achievements of generations past. Hyper-cooperative humans are willing to lose in the short term to win in the long term. Our remarkable ability to work together, on scales both large and small, has truly launched us to the top of the evolutionary ladder and continues to propel us forward into the unknown.

Hyper-cooperation isn't some preordained gift, however; it's a carefully cultivated skill, honed over millennia of shared struggle and triumph. As we face the challenges of the future, it's a skill we'll need more than ever before.

The Fallacy of Non-Zero-Sumness

Game theory assumes we can view human interaction as a game—and that we often subconsciously do this—and that our actions are just the ways we "play the game." But it goes beyond simple games; game theory is a powerful tool for understanding complex interactions between nations, businesses, and even societies as a whole. By analyzing potential outcomes and motivations, game theory helps us predict and understand how individuals and groups make decisions under various circumstances.

So often, we approach our world with a zero-sum mindset. In game theory, this means that there's an absolute limit of resources, and one person's loss is another's gain. Every win necessitates a corresponding loss, creating a limited framework for interaction. This is the mindset that says, "My company is stealing market share from my competitor: now I'll sell more products and they'll sell less." A zero-sum mindset says, "There are only so many seats in an entering freshman class: if some other kid gets a spot, my daughter won't." It's very easy to view the world in zero-sum terms because it allows us to claim winners and losers.

Americans are trapped in a zero-sum game every day when it comes to our national politics. With a political system overwhelmingly dominated by two parties, every seat lost by one is another "picked up" by the other. The U.S. Senate is a teeter-totter, and every seat that switches in an election could shift the entire balance of the country's politics. This is because it is a zero-sum situation: There are always one hundred Senate seats, and one side's loss is another's gain. The result of zero-sum situations like this is stagnation, conflict, and, as January 6, 2021, showed us, even violence.

In the end, the entire situation is truly zero-sum: the political machine of both parties grows and gets paid while the American people wonder if politicians on either side still represent their interests. We are all the losers. The problem is both the binary nature of the American political system and the mindset we bring to our elections: we, the people, encourage this zero-sum mindset, play into it, and perpetuate it.

But there are alternatives. I don't mean for our two-party system of politics in America (although perhaps we could do better there), but also alternative mindsets we can embrace as Americans—and as global citizens. Throughout his insightful work, Robert Wright has introduced a captivating concept: "non-zero-sumness." He explains:

> Both organic and human history involve the playing of ever-more-numerous, ever-larger, and ever-more-elaborate non-zero-sum games. It is the accumulation of these games—game upon game upon game—that constitutes the growth in biological and social

complexity that [other scholars] have talked about. I like to refer to this accumulation as an accumulation of "non-zero-sumness." Non-zero-sumness is a kind of potential—a potential for overall gain, or for overall loss, depending on how the game is played. . . . Non-zero-sumness, I'll argue, is something whose ongoing growth and ongoing fulfillment define the arrow of the history of life, from the primordial soup to the World Wide Web.[6]

This concept challenges the assumption that gains and losses must always balance out. Instead, Wright argues that interactions can create situations in which all parties involved benefit, exceeding the initial sum of what they had before. He applies the concept to our biological processes, our historical choices, and our destiny for the future, arguing that life that engages in non-zero-sumness has the evolutionary advantage in surviving long-term.

"Win-win" scenarios are those in which both parties benefit, creating a positive outcome that transcends the limitations of a zero-sum game in which one player's gain must equal another's loss. When we have chosen to win together, that is when our best ideas come to the fore. By exploring the principles of hyper-cooperation and win-win thinking, we can unlock the potential for mutually beneficial interactions, paving the way for a more prosperous and harmonious future rich with moonshot moments. We can also choose to build a world oriented toward sharing bright ideas, expanding wild thinking, and embracing moonshot moments, right at a time when we will need them the most.

So where do we begin in our discussion of the win-win approach? You would think that we'd begin with early man, but in fact, hyper-cooperation is not unique to our species. In fact, Tegmark argues that we embrace the computer science concept of "intelligent agents: entities that collect information about their environment from sensors and then process this information to decide how to act back on their environment."[7] By widening our definitions of who or what can win-win, we make space for the technologies we build to have actions and motivations of their own. But we also open a space to see how win-win scenarios are the foundation for the most successful life on our planet—and, most likely, for life we find in the cosmos as well. So let's begin with life at its simplest—and creepiest—to see how hyper-cooperation wrote the narrative of early evolution on our planet.

The Powerhouse Within Us

Any computer scientist will tell you that, if you want to understand how a computer system operates, you have to go dig into the code. Life is no different, and that code is very clearly written out in our chemistry as the As, Ts, Cs, and Gs. Life begins in the patterns of DNA, wrapped in helices trapped in our cell nuclei. These blueprints encoded the recipes for proteins, the formulas of our genes, and will become our paints in a synthetic biology future. But our cells, and the DNA within them, already demonstrate the most powerful example of hyper-cooperation in the living world—the mitogenome.

That's right—every one of your cells actually has two completely different sets of DNA, and one is just along for the ride.

Actually, that's not giving our mitochondria enough credit. These tiny power-houses are responsible for converting nutrients into energy, fueling everything from your thoughts to your heartbeat. They are in the cells of all living multicel-lular organisms, from plants to animals to fungi. But the mitochondria weren't always part of our cells.

According to the endosymbiotic theory,[8] billions of years ago, a primitive, single-celled organism engulfed a bacteria capable of harnessing energy through oxygen use. Instead of being digested, the bacteria formed a symbiotic relation-ship with its host, providing energy in exchange for a stable environment. Over time, this partnership became so beneficial that the two organisms merged, with the bacteria evolving into the mitochondria we know today.

This wasn't just a one-sided deal. The host cell gained a vital energy source while the bacteria gained protection and access to nutrients. This remarkable collaboration, born out of chance encounters and mutual benefit, marks a piv-otal moment in the evolution of complex life. It demonstrates how early cells, through a process of hyper-cooperation, could evolve beyond their individual limitations and create something far greater than the sum of their parts. Simply put: there is no life on our planet, at least as we know it, without a biological predilection for hyper-cooperation.

The Miracle of Mycelium

We can also see hyper-cooperation in some of the most primordial living beings on our planet: the vast network of mycelium networks thriving beneath our feet. These carpets of fungi produce mushrooms above the ground and exist as webs of intricate filaments weaving through the soil like silent threads, connecting millions of species of plants to a network where nutrients and information can be exchanged. The intricacy and symbiotic value of mycelium is a testament to the power of collaboration long before the rise of multicellular animals.

Fossil evidence suggests the presence of mycelium networks around 715–810 million years ago.[9] At that time, scientists saw a transition from single fungal spores, microscopic and alone, to more complex networks. The survival of that fungus hinges on finding suitable conditions to germinate and grow. But this is the magic of mycelium's hyper-cooperation: Upon germination, the spore produces delicate threads—tiny hyphae—that begin to explore the surrounding environment. These hyphae aren't solitary adventurers; they constantly communicate with one another, sending chemical signals and even fusing together to form a complex web—the mycelium network.[10]

This network is more than just a collection of threads; it's a marvel of information sharing and resource allocation. Let's posit a plant struggling with a nutrient deficiency. The mycelium network, with its vast reach, can detect this struggle. Through a sophisticated system of chemical exchange, the network can then transport vital nutrients from areas of plenty to the plant in need. This silent rescue mission is a testament to the interconnectedness and hyper-cooperative nature of the system.

The benefits extend far beyond nutrient exchange. Mycelium networks act as a communication highway for plants. Through this network, plants can warn one another of approaching danger, like a herbivore attack.[11] A plant being nibbled on by an insect can send a distress signal through the network, priming neighboring plants to release deterrents or even fortify their defenses.[12] This interspecies communication, facilitated by the hyper-cooperative mycelium, allows plants to collectively resist threats they might not overcome alone.

But perhaps the most compelling example of hyper-cooperation within mycelium networks lies in their role as decomposers. Dead organic matter litters the forest floor. Left alone, decomposition would be a slow, inefficient process. Enter the mycelium network, nature's ultimate recycling crew. Through the network's digestive enzymes, complex organic molecules are broken down into simpler forms, readily available for plants and other organisms to utilize.[13] This hyper-cooperation ensures the efficient return of nutrients to the ecosystem, a closed-loop system vital for sustained life.

The hyper-cooperative nature of mycelium networks isn't just about survival; it's about creating a thriving ecosystem. By fostering communication, facilitating resource exchange, and accelerating decomposition, these networks create an environment where all life benefits. It's a model of mutual aid, a silent symphony of cooperation playing out beneath the surface, and a testament to the power of collective action long before the first human handshake.

As scientists conduct further research into the world of mycology, they're increasingly recognizing the potential of these networks for bioremediation, pollution cleanup, and even the development of sustainable materials. In particular, mycelium's digestive enzymes break down a wide range of pollutants, including oil spills, plastics, and even pharmaceuticals.[14] Researchers are exploring ways to harness this ability for bioremediation, cleaning up contaminated environments effectively.[15] There are moonshot moments just creeping beneath our feet. By harnessing the power of these natural hyper-cooperators, we might unlock solutions to some of our most pressing environmental challenges.

The stories of mycelium networks and mitochondrial evolution are powerful reminders that the natural world is teeming with examples of hyper-cooperation. As we strive to build a more sustainable future, perhaps the wisdom of these silent collaborators within our bodies and beneath our feet can guide us. By learning to cooperate not just with one another but also with the intricate web of life that sustains us, we can pave the way for a future in which humans, like the mycelial networks, become not just individual entities, but vital parts of a thriving, hyper-cooperative ecosystem.

Teamwork Makes the Dream Work

Ants excel at hyper-cooperation because their very genetics have been fine-tuned for it over millions of years of evolution. While most of us carry a full set of genes capable of diverse functions, ant castes—workers, soldiers, queens—have specific genes selectively expressed, shaping their bodies and behaviors for specialized roles. Workers, for example, lack reproductive organs, dedicating their lives to foraging, brood care, and nest maintenance. This extreme specialization, unthinkable in most species, allows ants to function as a seamless superorganism, maximizing survival and resource acquisition.

Cooperation isn't just about common defense, it's also about sharing. Scarce resources demand efficient collection and storage, achieved through specialized workers dedicated to foraging and transportation. Unique reproductive strategies with specialized queens and synchronized breeding ensure colony survival even with individual losses, enabling exponential population growth. Finally, favorable genetic mutations promoting social behavior, sharing, cooperation, and communication were likely selected for over generations, laying the foundation for the complex social structures we see in ants today.

This brings us to a fascinating parallel with transhumanism and the ethical debates surrounding genetic engineering. If ants, through natural selection, achieved such efficient division of labor through genetic specialization, is it ethical for humans to manipulate our own genes for similar ends? Aldous Huxley's classic dystopia, *Brave New World*, explores this very question, presenting a society where individuals are genetically predetermined for specific roles, raising concerns about individuality and social hierarchies.

The key lies in finding a balance. Ants show us the power of hyper-cooperation and specialization, but their rigid caste system also carries drawbacks. Perhaps we can learn from them to embrace different skill sets and contributions while preserving individual freedoms and ethical considerations. As we navigate the complex landscape of transhumanism, studying the social insects, from the hyper-cooperative ant to the diverse honeybee hive, might offer valuable insights into maximizing human potential while safeguarding our ethical compass. Remember, while efficiency is desirable, we must avoid creating a society resembling an anthill at the cost of our humanity.

The Sweet Taste of Shared Success

Ants found success in part because of the evolution of flowers, a development that would change the very color of our planet.

About 130 million years ago, flowering plants, or angiosperms, differentiated themselves from their predecessors by evolving a dazzling secret weapon—nectar. This sugary treat, secreted by specialized structures called nectaries, marked the beginning of a remarkable story of hyper-cooperation between plants and pollinators.[16]

As pollinators sip nectar, they inadvertently collect pollen on their bodies, transferring it from flower to flower, enabling fertilization and plant reproduction. This mutually beneficial relationship—pollination—proved to be a winning strategy for both parties. This explosive success story stands as a testament to the power of hyper-cooperation in evolution, again and again.

For the first 20 million years or so, flowering plants slowly gained traction. But around 100 million years ago, a colony of ants figured out a win-win relationship with flowers, grew some wings, and evolved into bees.[17] Around that same time, moths looking to escape competition for nighttime feeding started feeding on nectar during the day, evolving brighter colors and unique characteristics to become butterflies. Through these co-evolutions, the hyper-cooperative fate of flowers was inevitable. Flowering plants diversified rapidly, evolving into a dominant form of plant life on Earth with more than 350,000 species today.

Just as the delicate dance between flowering plants and pollinators sparked a co-evolutionary arms race millions of years ago, our development of AI holds the potential for a similar, hyper-cooperative future. Flowering plants, with neither intentionality nor consciousness, utilized the principle of hyper-cooperation to evolve a viable food source, and then essentially waited until other organisms could co-evolve enough to cooperate and benefit from it. Flowers created a situation where a moonshot moment could occur (for both bees and butterflies in fact), leading to a win-win for all.

If flowering plants could do this, then so can I, and you, and all of us. Right now, I bet you feel a little weird reading "if flowers can do it, so can we!" Perhaps it feels like I'm anthropomorphizing the roses and daisies. But that has been the point of this entire discussion: to illustrate that the basic nature of hyper-cooperation is so powerful that it does not require consciousness to choose it. Life moves toward hyper-cooperation when it wants to succeed, in part because hyper-cooperation creates the opportunities for wild change and unprecedented problem-solving.

When I think of this, I feel reassured. If consciousness is not required to find one's way to hyper-cooperative thinking, then there is a real chance that we can steer AI down a path of hyper-cooperation with humanity if that's what we choose to do. We won't be pushing against natural tendencies: We won't be asking AI to disregard the laws of the universe. As long as we encourage AI to follow the path of the mitochondria, the mycelium network, the ant, and the flower, we might have a chance at a truly bright future.

Becoming Us

Let me tell you a story about humanity and fire. It might be the greatest story of all time.

First, let me set the scene: 800,000 years ago, a bone-chilling wind whips across the vast African savannah. Jagged shadows from acacia trees stretch across the cracked earth, painting an endless dance on the rusty horizon. We, early humans, huddle within the cool embrace of our cave dwelling, carved into the edge of this unforgiving landscape. Though our brains are still evolving, a spark of intelligence shines brighter within us than in any other ape. We stand tall on two legs, the legacy of bipedalism etched onto our frames. No longer bound to the trees, we venture out onto the open plains, clutching our crudely sharpened sticks, the tools of our nascent hunting prowess. Life remains precarious, each day a delicate balance between meager survival and the constant threat of danger.

We are not unique. There have been other proto-humans. Some have already walked off the continent and made their way to Europe and Asia. Many have evolved specialized teeth or feet; some have already disappeared to evolution's pressures. Indeed, we are not at the top of our local food chain. In fact, we've traded our strength and speed compared to other apes for bigger brains that are hard to feed and not doing much for us beyond those pointy sticks.

But around one million to 800,000 years ago, early humans figured out a very important tool: fire.[18] Fire meant a source of heat and light, a way to scare off predators and make a wider range of shelters possible. But fire is, oddly enough, a highly cooperative enterprise. That's because it is far easier to keep fire than to make it.

Before early humans learned the complex techniques of fire-making, flickering flames were fleeting gifts snatched from nature's grasp—a lightning strike setting the dry grass ablaze, smoldering embers left by volcanic activity, or perhaps even the remnants of a previous forest fire.[19] Keeping these flames alive wouldn't have been easy. To ensure warmth and protection continues, fire needs constant attention, something blocking it from the elements, and someone to feed it new fuel. This constant vigilance was a heavy burden for a single individual, but sharing the responsibility within a group offered significant advantages. Everyone can work cooperatively to maintain a communal fire, and it's a resource that is easy to share because it is not fixed.

Fire is a win-win for groups—well, unless things start burning down, of course. Then again, it's also easier to put out a fire if you have some spare hands to help.

Fire also let early humans do something that makes us so unique: cook our food.[20] While the spark that ignited this culinary journey remains lost to time, the possibilities intrigue us. Perhaps a stray piece of meat fell into the flames, revealing its transformed, tender texture and delicious aroma. Maybe someone dropped a root vegetable too close to the embers, witnessing its surprising transformation into a softer, more palatable form.[21] Whatever the initial discovery was a million years or so ago, the implications were profound.

Our precooking ancestors spent their days dominated by scavenging and chewing tough, raw food. Digesting these meals was a time-consuming, energy-intensive chore, requiring bulky intestines to break down the fibrous material or bacteria-ridden meat. Enter the game-changer: cooked food. The application of heat softened meat and vegetables, making them easier to digest. This culinary revolution, driven by chance or experimentation, had a profound impact on our evolution. As our bodies no longer needed massive digestive systems for raw food, our intestines could shrink.

This freed up a significant amount of energy, which, according to Aiello and Wheeler's Expensive Tissue Hypothesis,[22] was then redirected toward nourishing larger and more powerful brains. Cooking food, from meat to tubers to nuts and grasses, made our brains bigger and our minds smarter. This theory, while not without its critics, offers a compelling explanation for how the simple act of cooking our food fueled the growth of our cognitive abilities, propelling us on the path toward becoming the intelligent beings we are today.

Cooking presented our ancestors with many new challenges: gathering suitable food, tending the fire, experimenting with cooking methods, and ensuring everyone received their share. No single individual could manage this complex process alone. Instead, it demanded the development of hyper-collaboration. One person might gather firewood while another tended the flames. Others might experiment with different foods, sharing their discoveries and failures.[23] The knowledge gained, the skills honed, and the food itself were all communal resources, shared within the group. This hyper-cooperation wasn't just about survival; it fostered social bonds, communication, and a sense of shared responsibility.

Around 400,000 years ago, early humans added another cooperative tool to the box: big-game hunting.[24] Big-game hunting wasn't a solo adventure for early humans; it was a strategic, hyper-cooperative affair fueled by fire. We're talking mammoths, rhinos, even giant deer, which are not only large but also congregate in groups and herds. Taking down such behemoths required collaboration beyond brute force.

Big game required bold thinking.

The hunt itself was a ballet of coordinated effort. Skilled runners would herd the target, using fire strategically to create barriers or stampede the animal toward waiting companions armed with spears and sharpened stones. Hunts would take several days, requiring a mastery of fire to stay warm while away from home for days at a time. Fire also played a crucial role in post-kill processing. Not only did it ward off scavengers and provide warmth, but it also allowed for butchering and preserving meat, ensuring everyone in the group benefited from the shared bounty.

This hyper-cooperation wasn't just about efficiency; it was about survival. Facing a colossal creature alone was a recipe for disaster. But by working together, pooling their skills and resources, early humans could overcome these seemingly insurmountable challenges. The rewards were immense: not just meat, but also bones for tools, hides for clothing, and a sense of collective achievement that strengthened social bonds within the group. Big-game hunting was a crucible for forging our ancestors' resilience and resourcefulness, all thanks to the power of hyper-cooperation fueled by the ancient magic of fire.

With big game in their bellies, hides for carrying food and water great distances, and fire to keep them warm, early man ventured out from Africa multiple times, battling Neanderthals and others along the way. Our large brains, fueled by roasted venison and softened carrots, could solve problems and handle communication at a level that no other species on the planet could. By 70,000 years ago, a cognitive revolution had begun, energized by fire, food, and hyper-cooperation.

That cognitive revolution brought us the last great invention, the one that made us truly human and set *Homo sapiens* apart from all the others, once and for all. It wasn't tools—we'd had those for over a million years, and so do other species like crows and monkeys. It wasn't fire, since we know Neanderthals and others conquered that. It wasn't even language, which was the moonshot moment of our cognitive revolution. After all, lots of animals communicate through language of various forms, and it's fairly clear we developed language as a means to an end, to express ideas we already wanted to share.[25]

It was that story itself.

Around 30,000 years ago, humans began painting on cave walls and carving figurines of anthropomorphic animals. During this period, archaeologists believe that early humans developed the concept of myth.[26] Diverse lines of evidence, including cave paintings in Lascaux and Chauvet,[27] Venus figurines made concurrently in unrelated locations around the world, and early stone carvings in the Middle East collectively point to a shift in human cognition when our ancestors were no longer just surviving; they were actively seeking meaning and explaining their world through stories and symbols.

Indeed, it is storytelling that makes humans unique: the opportunity to share knowledge, pass on values, and remember ancestors. No other animal tells tales of its past. No other animal expands its minds to such openness. Myth requires a huge cognitive leap: you have to talk about things that have passed, that you were not even witness to; sometimes, you have to talk about things to come, or futures that would have happened if not for the hero. This ability to use language in such a way is what makes us unique.

Obviously, myths are a hyper-cooperative invention. Myths allow for the passing of knowledge, which is essential for humans as an intelligence based species. Myths are shared between members of a community, bringing people together and creating a shared identity.[28] In fact, myth is a cooperation-multiplier: We cooperate to create myths, and then they tie us to communities that then cooperate and build other things. As Harari puts it, "any large-scale human-cooperation—whether a modern state, a medieval church, an ancient city, or an ancient tribe—is rooted in common myths that exist only in people's collective imagination."[29] In the end, the technology that ties us together, as hyper-cooperative beings, is the shared myth.

Our first shared myths were told by the light of a fire while listeners sat, enraptured, chewing on juicy roast meat and feeling the strength of their cooperative success. So that's where we'll close our story of humanity and fire: in the end, it wasn't the fire that was most important, but the stories we told (and have kept telling) around it.

Domesticating Plants, Animals, and Ourselves

We have now walked through early human history, seeing how our biology pushed us toward hyper-cooperation and sizzling mammoth steaks. Joseph Henrich argues humans regularly take these new ideas and turn them into codified knowledge that we share, generation after generation.[30] New hunting techniques, knowing where to harvest certain plants or fungi—these are all ways that culture perpetuates itself through knowledge sharing. In this same way, domestication of plants and animals marked a huge leap in survival but also marked a cultural revolution: one which ultimately allowed humans to domesticate themselves.

Agriculture was one of those moonshot technologies that developed independently in multiple parts of the world around 12,000 years ago. Taming the chaos of nature and coaxing seeds to sprout in controlled environments was no easy feat. It demanded collaboration and was fraught with uncertainty.[31] Groups of individuals, sharing knowledge and resources, observed, experimented, and learned from failures. The very act of planting, tending, and harvesting required coordinated effort, a far cry from the solitary and small group hunter-gatherer lifestyle.

Furthermore, agriculture wasn't just about feeding oneself; it was about creating surpluses, about supporting those who couldn't farm themselves. This act of sharing, of ensuring the collective well-being above individual needs, is the very essence of hyper-cooperation. It allowed for specialization, for the rise of artisans and scholars who could flourish thanks to the steady food supply secured by farmers.

But our hyper-cooperative spirit didn't stop at plants. The domestication of animals like sheep, goats, and cattle was another act of partnership. Around 10,000 years ago, humans and dogs domesticated each other, finding hunting and herding partners, companions, and friends. This interspecies cooperation, forging bonds across biological boundaries, further cemented the power of collaboration in shaping our evolution.

We often talk about agriculture and animal husbandry—domesticating plants and animals—as technological leaps, but we often forget that it was less about leaping and more about settling down. With crops and herds, we domesticated ourselves. The shift from nomadic lifestyles to settled communities around 8,000 years ago, driven by the stability of agriculture, marked a dramatic change in understanding what it meant to be human. Our genetic codes changed, making it easier to digest foods like cows' milk and potatoes.[32] Cities rose, cradles of civilization, teeming with diverse individuals working toward both unique and shared goals. This new reality demanded new forms of hyper-cooperation.

Shared belief systems like religion emerged, providing moral frameworks and social cohesion. Political structures were established to manage complex societies, ensuring order and justice. Sanitation systems were developed, demonstrating the collective responsibility for public health. And of course, complex food distribution networks arose, feeding the growing urban populations.

This domestication of humans wasn't about taming or controlling; it was about adapting, evolving, and finding new ways to collaborate in the face of unprecedented challenges, and using our predisposition toward hyper-cooperation to expand the species. It wasn't always smooth; conflict and inequality remain scars on our history. But the success of the underlying principle—the ability to work together toward shared goals—is undeniable.

Looking back, we see a clear pattern: domestication, whether of plants, animals, or ourselves, has always been a hyper-cooperative endeavor. Will we seek to domesticate AI, or robots, or our genetic creations? Likely, yes, but domestication can be a win-win for all. Our agricultural past is a testament to the power of collective action, of pooling resources, knowledge, and effort to achieve what individuals cannot. This legacy extends far beyond our early farms and herding dogs. From the collaborative spirit that fuels scientific breakthroughs to the international treaties that address global challenges, we continue to harness the power of hyper-cooperation, building upon the foundation laid millennia ago.

Harnessing Moonshot Momentum

When groups of organisms, including humans, hyper-cooperate, they transform into something far greater than the sum of their parts, creating fertile ground for those seemingly impossible breakthroughs we call moonshot moments.

So what is it about hyper-cooperation that encourages moonshot moments?

Hyper-cooperation doesn't just leverage individual strengths; it also pools resources and shares risks. The mammoth hunt wouldn't have been possible without everyone contributing their tools, skills, and even the willingness to face potential danger. This risk-sharing empowered them to attempt grander endeavors, knowing that failure wouldn't spell individual doom, but a shared learning experience for the group. It's a principle that transcends time, from those early hunters to the collaborative efforts behind the moon landing, when the collective will and shared resources of countless individuals propelled humanity to achieve the unthinkable.

But what truly fuels these moonshot moments is the powerful momentum that arises from hyper-cooperation. You can imagine an early woman who has figured out that moonshot idea, "cook the plant before you eat it," working her way through an entire collection of tubers, herbs, and leaves while her daughters and sisters make a mental catalog of tastes and effects. This same synergy is evident in the collaborative environments that birthed groundbreaking inventions, from the bustling workshops of the Renaissance to the modern-day tech giants, where shared goals and a culture of collaboration fuel innovation at breakneck speed.

Finally, hyper-cooperation allows groups to think beyond immediate needs and set their sights on the long horizon. The early humans weren't just hunting for a single meal; they were ensuring the survival of their community, of future generations. This shared vision, this ability to make sacrifices for a future they might not even see, is what empowers groups to achieve the seemingly impossible. It's the thread that connects the dots between the mammoth hunt and the hunt for a cure to Alzheimer's, between the collaborative narratives woven around campfires and the international partnerships of today that tackle global challenges.

Hyper-cooperation isn't just about efficiency; it's about unlocking the collective potential for moonshot moments. By amplifying intelligence, sharing resources and risks, and setting long-term goals, groups can achieve what individuals cannot, leaving their mark on the world in extraordinary ways. From the echoes of our ancestors' shared fires and farms to the collaborative endeavors that shape our future, the power of hyper-cooperation continues to illuminate the path toward the seemingly impossible.

Paradigms for Hyper-Cooperative AI: Life 3.0 and the Singularity

Before moving forward, I want to explore a view of society built around hyper-cooperativity between not only ourselves, but with AI: Max Tegmark's "Life 3.0" paradigm and the enigmatic concept of the technological singularity. I want to compare each of these unique perspectives offered, in part because they relate to our earlier discussions of choice, and because both are inherently predicated on the idea of hyper-cooperation. "Life 3.0" proposes a gradual yet transformative merger of human and artificial intelligence, culminating in a collective intelligence exceeding our current boundaries. The singularity, on the other hand, posits a more abrupt and potentially unpredictable shift, a tipping point when intelligence explodes beyond our comprehension. By comparing these contrasting viewpoints, we can gain a richer understanding of the potential pathways our hyper-cooperative future might unfold.

Max Tegmark, a renowned physicist who has been a player in the AI game for decades, proposes a provocative framework for understanding the evolution of life: Life 1.0, 2.0, and 3.0.[33] Conveniently labeled like a new software release, each stage represents a distinct level of complexity and intelligence.

Tegmark argues that "Life 1.0" encompasses biological life based on organic molecules and natural selection. This includes everything from bacteria to the most complex animals, including ourselves in our current state. Our intelligence here arises from biological processes within individual organisms. He describes this as a stage where biology controls the hardware (our bodies) and the software (our behaviors).

"Life 2.0," where we currently stand, introduces cultural evolution. We possess the unique ability to learn, share knowledge, and build upon the achievements of previous generations. As we've discussed in this chapter, this collective intelligence, fueled by language, tools, and technology, sets us apart from other animals. In this next phase, our biology still controls our hardware, but we write our own software, through cultural norms and practices.

However, Tegmark suggests that "Life 3.0" represents a potential quantum leap where artificial intelligence surpasses human intelligence and merges with our biological existence. This could involve direct brain-computer interfaces, AI-powered augmentation, or even complete mental uploading. While the specifics remain speculative, the core idea is a transformative integration of biological and artificial intelligence: the next step in hyper-cooperation.

Tegmark's vision of Life 3.0 is both exciting and challenging. It carries the potential for unprecedented advancements in various fields, but Tegmark also raises critical ethical and philosophical questions about individual identity, control, and what it means to be human, some of which I will address later in this book. Tegmark isn't naive; he offers plenty of warning against the dangers of Life 3.0 but argues that it is necessary to align the goals of AI with our own to avoid the worst possibilities of AI development. Tegmark's Life 3.0 paradigm offers one vision of humanity's hyper-cooperative, win-win nature being extended to the next logical step in regard to AI development.

In comparison, the concept of the singularity implies far less human control over the course of technological events. The *singularity*, a term now ingrained in the lexicon of futurism, describes a hypothetical moment in time characterized by a rapid and irreversible acceleration of technological progress, driven by artificial intelligence (AI). Beyond that point, our current understanding of technology, society, and even human existence itself is said to become obsolete. The implications of this concept are enormous, encompassing both utopian dreams of boundless potential and dystopian nightmares of existential threats.

The term *technological singularity* was first popularized by computer scientist and science fiction writer Vernor Vinge in his 1993 paper "The Coming Technological Singularity." Here, he described the singularity as "a hypothetical future event at which technological growth becomes uncontrollable and irreversible, resulting in unforeseeable consequences for human civilization."

Vinge didn't offer a specific date for this tipping point, but others have taken up the challenge of prediction. Renowned inventor and futurist Ray Kurzweil famously predicts the singularity around 2045, propelled by rapid advancements in AI, nanotechnology, and biotechnology.[34] He envisions a future in which humans and machines merge, transcending biological limitations and ushering in an era of unimaginable growth and prosperity.

However, not everyone shares Kurzweil's optimism. Philosopher Nick Bostrom, known for his work on existential risk, AI, and transhumanism, argues that the singularity could pose major risks.[35] He highlights the possibility of superintelligent AI surpassing human control, leading to unintended consequences or even existential threats. Bostrom urges us to approach the singularity with caution and develop "friendly AI" to ensure its alignment with human values. Similarly, Vinge has expressed concerns about the potential dangers of the singularity, stating in his 2006 essay "The Long Reflection" that it's a "dangerous crossroads" and humanity needs to be prepared for unforeseen consequences.

So, are we currently hurtling toward the singularity? There's no definitive answer. While we're witnessing impressive advancements in AI and related fields, it's difficult to predict the exact pace and trajectory of technological progress. The singularity might be decades away, or it might never happen at all. And I tend to agree with Mustafa Suleyman when he writes, "I believe the debate about whether and when the Singularity will be achieved is a colossal red herring."[36] When we will reach the singularity, or how, is not as important as the practicalities of the tech we are building right now, and the ethics and values that the singularity represents that are worth applying today.

So what can we draw from both Tegmark's Life 3.0 paradigm and the broader concept of the singularity? I'd argue that both point to a real need to focus on building what Eliezer Yudkowsky, founder of the Machine Intelligence Research

Institute (MIRI), calls friendly AI—AI that has its goals aligned with ours and remains under our control or at least in a hyper-cooperative state.

While the seeds of the concept of friendly AI can be traced back to I. J. Good's seminal 1965 paper "Intelligence Explosions and Orbits," in which he warned of the potential dangers of unaligned superintelligence and advocated for "benevolent superintelligence," it was Yudkowsky who truly crystallized the term in 2001. In his discussion of friendly AI, Yudkowsky tackled the technical and philosophical challenges of aligning AI goals with human values, proposing frameworks for "recursive self-improvement" to understand the nature of AI ethics.

Both Bostrom and Tegmark have built upon this concept. Bostrom warns that creating powerful AI without proper alignment could be catastrophic, citing scenarios where an AI, optimized for efficiency, might misinterpret and harm humanity in its pursuit of a goal like maximizing energy production. He cites the "value drift"[37] problem, where even well-intentioned AI could diverge from human values over time, potentially leading to unforeseen negative consequences. Bostrom also argues for "orthogonality"—the idea that intelligence is a tool, independent of morality and can be used for good or bad: it thus behooves us to build friendly AIs, because it is the friendliness, not the intelligence, that determines the moral character of that AI.

Tegmark argues that friendly AI wouldn't be achieved through mere coding or safeguards, but through fostering a "shared moral compass" between humans and AI.[38] Ultimately, Yudkowsky, Bostrom, and Tegmark all argue that simply hoping for benevolent AI won't suffice; we need proactive strategies to ensure AI alignment with human values and goals. It is not that we should fear superintelligence outsmarting humanity; rather, we should ensure we don't give AI access to too much power too quickly. As Ben Kuipers joked to me during a chat, in our worst sci-fi visions of malicious AI, "it's not that those AI are super intelligent, though they are very clever. It's that they are incredibly powerful. The mistake is giving them the keys to the nuclear weapons. So that's the mistake we gotta not make." Ensuring AI is friendly is an exercise in self-restraint.

The concept of friendly AI goes beyond just training benevolent intelligence; it's a call to action for a new paradigm of technological development. I believe we can take a step back from both the Life 3.0 approach and the question of the singularity and instead imagine not just friendly AI designed to serve humanity but all emerging technologies, from synthetic biology to robots, built with the same guiding principle: alignment with human evolution and shared prosperity. This, in essence, captures the heart of a win-win, hyper-cooperative vision for the future of tech, one ripe for the blossoming of many moonshot moments.

But how do we translate this aspirational goal into reality? How do we pursue a win-win future, building a greater human community oriented toward hyper-cooperation while also ensuring a collaborative relationship with the technologies we make? I believe the answer lies in international cooperation and collaboration between academia, industry, and governments to first iden-tify a clear trajectory for our work: defining our species' collective technomoral virtues and determining our technogoals and priorities.

The First Step: Defining Our Virtues and Goals

If all future regulation, innovation, and development is predicated on our ability to figure out what it is we "want" as hyper-collaborators, then we need to work toward forging shared values and aligning toward collective goals.

A prescription of forging shared values feels naive. I'd hope we could all agree on universal human rights. But clearly, we do not. Shared values surrounding battling climate change? That's practically a political third rail. Surely, though, we can all agree that biochemical warfare should be banned? Or that the media should not post fake news or artificially produced images?

There have been experiences of global cooperation. In fact, the United Nations Sustainable Development Goals (SDGs) stand as a beacon of hyper-cooperation on a global scale. Adopted in 2015 by all UN member states, these seventeen interconnected goals represent a collective vision for a better future for people and the planet.[39] From eradicating poverty and hunger to ensuring quality education and clean water for all, the SDGs tackle some of humanity's most pressing challenges.

This ambitious agenda hinges on collaboration between governments, businesses, and civil society. We've seen this cooperation bear fruit in various ways. Extreme poverty rates have been halved since 1990, with millions lifted out of economic hardship. More children are now enrolled in primary school than ever before, and access to safe drinking water has significantly improved. The SDGs may be a long-term vision, but the progress made so far demonstrates the power of hyper-cooperation in creating a more just and sustainable world for all.

The road toward achieving the SDGs hasn't been without its bumps. Progress has been uneven across regions, with some countries facing greater difficulties due to conflict, poverty, or limited resources. The COVID-19 pandemic was a significant blow, pushing millions back into poverty and jeopardizing years of progress on health and education goals. Additionally, rising inequality and climate change pose constant threats, jeopardizing the gains made. Fulfilling the ambitious vision of the SDGs requires sustained commitment, innovative solutions, and continued global cooperation, especially in the face of these ongoing challenges.

Thus, humanity has a mixed track record when it comes to establishing a code of shared values, yet when it comes to our approach to the coming wave of inevitable technological change, we will need to create a unified ethical framework for working with that technology.

Shannon Vallor offers a novel approach to ethical technology development: the concept of technomoral virtues.[40] She argues that our existing frameworks, often rooted in legal regulations or abstract principles, fall short in the dynamic landscape of digital advancements.

Instead, Vallor proposes cultivating specific character traits within individuals and societies that guide our interactions with technology and its consequences. These "technomoral virtues" go beyond mere technical expertise, encompassing essential values like humility, acknowledging the limitations of technology and ourselves; justice, ensuring equitable access to technology's benefits; and care, actively using technology to address societal needs. This virtue-based approach offers a refreshing perspective, encouraging conscious development of ethical character as a cornerstone of responsible technology development and use.

Rather than rigid rules, codifying technomoral virtues demands a shared moral compass cultivated within the tech community, in collaboration with academia, national governments, and international organizations. Scholarly exchanges, real-world case studies, open-source resources, educational integration, and recognition programs would allow humanity to collectively define, refine, and disseminate these crucial values. This interactive process would allow humanity's technomoral virtue system to stay dynamic and relevant, ensuring a tech landscape steered by shared ethics and aspirational principles, not static codes.

A final note on determining humanity's shared technomoral virtues: some may consider the definition and refinement of a shared set of morals to be an unnecessary step, or a purely academic exercise. However, it was, as I explained earlier in this chapter, the very process of defining our values and codifying them as myths and shared narratives that separated humans from other species and pushed us to become the dominant species on this planet.

And if we ever reach the point where we create AI with superintelligence, or genetically-engineered progeny that are no longer fully human, or robots with dreams of their own—don't we want to be able to say we were steered by some compass, that we had some set of ethics guiding us to good?

Don't we owe it to future generations to carve out a moral high ground if they ever need refuge from the coming wave?

Building a hyper-cooperative future also hinges on shared "technogoals." Working toward shared goals is humanity's greatest challenge, and goals are "what the thorniest AI controversies are about."[41] Without a clear, shared vision, our technological advancements risk becoming fragmented, misaligned, or even harmful.

Therefore, our focus should be on defining what I would call technogoals—goals specifically oriented toward the development and deployment of advanced technology. These goals shouldn't be abstract aspirations; they should be concrete, measurable objectives that reflect our collective priorities. Eradicating major diseases, maximizing agricultural efficiency, and mitigating climate change are just a few examples of various obvious technogoals we could set at the start.

By codifying these goals through dedicated bodies, official documents, and robust regulations, we create a tangible framework for cooperation. We need to establish a shared narrative, a collective ambition that unites individuals and nations in a common purpose.

While the notion of shared technogoals is undeniably attractive, achieving them presents a formidable challenge. The very forces driving technological advancement can work against us. Consider the immense profits potentially accruing to companies at the forefront of breakthroughs like AI or genetic engineering. Such financial incentives can make it difficult to coax industry leaders to the table for collaborative goal setting, especially if those goals deviate from maximizing shareholder value. As Mustafa Suleyman aptly argues, containing the potential risks of emerging technologies requires a unified set of goals, yet the current system often incentivizes a different path.

Therefore, we might need to nurture a new breed of entrepreneurs—those who prioritize shared goals and social good over pure profit maximization. This necessitates a shift toward "purposeful capitalism," where businesses demonstrably align their activities with the betterment of society. While this transformation won't be easy, it's a crucial step toward establishing shared technogoals. Remember, building a hyper-cooperative future will require not just technological innovation but also a fundamental shift in our economic and societal mindset.

But the benefits extend beyond mere coordination. Shared technogoals also ensure that, as we develop AI and other transformative technologies, these innovations are aligned with our values and aspirations. In Tegmark's words, "the real risk with AGI isn't malice but competence. A superintelligent AI will be extremely good at accomplishing its goals, and if those goals are not aligned with ours, we're in trouble."[42] This means building safeguards, embedding ethical considerations into the very fabric of our technological advancements, and ensuring they serve the betterment of humanity, not its destruction.

Ultimately, embracing shared technomoral virtues and technogoals is not just an exercise in idealism; it's a strategic imperative for harnessing the immense potential of technology while mitigating its risks. By fostering a global culture of hyper-cooperation, guided by a clear set of shared objectives, we can navigate the technological landscape with greater wisdom and purpose, ensuring that our moonshot moments truly propel us toward a brighter future for all.

Planting the Seeds of Hyper-Cooperation

The road to a hyper-cooperative future may seem daunting in a world of polarized elections, simmering international tensions, and the ever-present allure of individual financial gain. Yet, as Mustafa Suleyman argues, "if we—we humanity—can change the context with a surge of committed new movements, businesses, and governments, with revised incentives, with boosted technical capacities, knowledge, and safeguards, then we can create conditions for setting off down that teetering path with a spark of hope."[43] While the current political climate presents undeniable challenges, despair is not an option. We must find the seeds of hyper-cooperation wherever they may sprout.

Perhaps it's time for all of us to encourage the business world, driven by its global reach and innovative spirit, to join hands with academia, the wellspring of knowledge and critical thinking. This unlikely partnership, fueled by shared technogoals and a commitment to the betterment of humanity, could become the catalyst for a global shift in mindset. Everyday citizens must also encourage governments to get involved in scientific development[44] and to work toward multinational agreements on ethics, regulations, limits, and more. It's not going to be easy, but we will not foster a moonshot moment mindset without an international commitment to hyper-cooperation. We all have to win-win together.

While the vision of a hyper-cooperative global society primed for moonshot moments might seem utopian, remember, even the most bold journeys begin with a single step. While achieving this ideal may ultimately require systemic shifts, individual action serves as the fertile ground from which change can blossom. This begs the question: How can we, as individuals, cultivate the read-iness and openness necessary to embrace moonshot moments within our own spheres of influence? If we cannot push our society into hyper-cooperation right now, can we as individuals plant these ideas, and let them spread like that slime mold, on the path of max efficiency to a brighter future for all?

PART III

PREPARING THE SELF FOR GREATER CREATIVITY

Chapter Six

ASSOCIATIONAL THINKING AND PERSONAL KNOWLEDGE MANAGEMENT

Ever since I was a child, I have been blessed (or perhaps cursed) with insatiable curiosity. Even before I reached middle school, every book, story, or documentary I devoured left a trail of sloppy notes—a physical manifestation of my overflowing mental space. Ideas, like fireflies, often flitted through my mind, each one captivating me for a few weeks before I chased the next glimmer.

This free ranging mind wasn't always ideal. As the child of immigrants, new to the United States, I struggled with switching up my Farsi for English. My language skills were so bad I became fixated on the idea that I must be stupid, or dyslexic, or incapable of processing language. My teachers added to the anxiety. I remember them leaving notes for my mother on my report cards.

- "Milan is always daydreaming and can't focus."
- "Milan is careless in his work and lacks attention to detail."
- "Milan struggles to pay attention in class."

These felt like a damning indictment, restrictions on my ability to reach my full potential as well as my ability to be organized. They ignited a long-standing internal conflict within me—a desire for order amid the chaos, a means of channeling my curiosity without giving in to its disorder.

Middle school was a time marked by both the awkwardness of adolescence and the heartbreaking decline of my grandmother's mind. Witnessing her struggle with Alzheimer's and dementia left an indelible mark. The gradual erosion of her ability to recognize loved ones, the frustration in her eyes as familiar words slipped away, the fading spark of connection—it was a brutal education on the fragility of memory and the human condition. This experience, though deeply painful, became a powerful motivator. It ignited a passion within me to "hack" my own mind, to build a system of knowledge management that wouldn't just store ideas but would actively protect them. The fear of losing myself, or someone I love, to the ravages of time became a driving force in my quest to understand and fortify my own mental landscape.

As I anticipated, digital tools would finally provide me the framework to order my thoughts and tame the imagined scatterbrain within, technology suddenly became a potential weapon in this struggle. But throughout college, my personal knowledge management (PKM) system wasn't a system at all, but a chaotic dance between Dropbox folders, overflowing Google Docs, and a to-do list app perpetually teetering on meltdown. My notes lacked a centralized categorization system and instead were filed according to a mess of personal codes I alone understood. It functioned, after a fashion, but the friction was undeniable.

Then, I stumbled upon Notion, especially through the work of August Bradley. It felt like a revelation. The sheer versatility of the platform was mind-blowing. A single page could morph into a project management hub, a comprehensive wiki, even a sleek landing page. The nesting capabilities allowed for organization within organization, and the customization options felt like having a personal digital Swiss Army Knife. Databases—an entire universe of information housed within a single, searchable landscape—were the icing on the cake.

So, while writing my last book and running multiple startups was a whirlwind a couple of years ago, I built myself a secret weapon—a system for managing the information fire hose. Think meticulously organized digital notes, voice memos buzzing with post-talk insights, and transcripts of interviews—a personal knowledge vault of all my research waiting to be tapped, all organized in my PKM system.

The thing I love the most about my PKM system is its comprehensiveness. My vaults are more than a repository for grand ideas and project plans: they serve as a digital scrapbook of my life. Nestled within their digital walls are the heartfelt words of my sister's wedding speech, a poignant archive of letters exchanged with my parents over the years, even the raw, emotional outpourings of past breakups. My system now holds my health records, doctor's appointment notes, all my findings from my biohacking journey I'll walk you through in Chapter Eight, even my genetic test results—a detailed chronicle of my body's default modes of operation.

Notion's capacity to function as a top PKM tool underwent a substantial transformation in 2024, which has significantly aided me in making further use of the years of information I have catalogued. With the help of artificial intelligence, Notion was able to evolve from a robust organizer to an almost sentient personal assistant. My personal data was used to train my own personal Large Language Model (LLM) that I can use for many different purposes within Notion. I recall the early Notion adoption days, when we users would half jokingly send bug reports and feature requests with a gleam in our eyes, expecting for just such an evolution. It's a reality now. This artificial intelligence, this extension of me made from my data, is a powerful means of self-improvement. It's a glimpse into the possibilities of early transhumanism, albeit from a consciousness-enhancing perspective within the digital realm.

In retrospect, the seeds of my PKM journey were sown in that fertile ground of childhood disorganization: I did lack focus, honestly. The me of yesteryear, buried under a disorganized maze of random paperwork and undeveloped concepts, struggled with a persistent nagging fear of handling the knowledge that was all around him. Fortunately, PKM showed up as a lifesaver—an organized method that managed the flood of data while simultaneously encouraging a more narrowly focused way of thinking. With the help of this system, I've been able to overcome information overload and bring my dreams—including the book you're reading right now—to life.

Getting to Know PKM

Every day, a tsunami of articles, emails, podcasts, and social media snippets crashes over us, threatening to drown our ability to focus and think clearly. Yet, amidst this constant, chaotic deluge lie powerful opportunities—fleeting moments of inspiration, connections that spark innovation, ideas with the potential to change the world. These crystalize as moonshot moments, the sparks that ignite breakthroughs and progress.

But how do we capture these fleeting insights before they vanish in the digital downpour? And how do we cultivate an environment where inspiration can take root and flourish? We have already discussed societies more focused on hyper-cooperation, but how do we then foster such win-win cooperation within our own minds? We can begin upgrading ourselves to handle this flood of data by embracing personal knowledge management. I tend to think of PKM not as building a "second brain," a cold, sterile storage locker for facts, but as a vibrant knowledge oasis to play in. It's a place where the seeds of ideas are planted, nurtured, and allowed to grow into something remarkable.

A PKM system is an overarching organizational mindset, digitally laid out in a data-management application, to make sorting, storing, and associating random notes easier. It is essentially a way to upgrade yourself, without implants or complex tech. You really only need a cell phone or desktop, making PKM accessible to anyone looking to level themselves up right now.

Here's the magic of PKM: it doesn't just help us store information; it unlocks its potential. By actively collecting, organizing, and reflecting on the information we encounter, PKM empowers us to transform the information overload from a burden into a springboard for creativity and innovation.[1]

Think about the last time you bumped into a stellar article with lots of thought-provoking ideas, but didn't know where to put the tidbits you'd gathered—and a day later, you'd sadly forgotten. With a robust PKM system, you can not only save the article but also connect it to past notes on similar topics, flag key takeaways, and explore potential applications. This act of curation and reflection transforms the article from a fleeting blip on the information radar into a vital piece of your personal knowledge ecosystem.

The benefits of PKM extend far beyond mere organization. By actively engaging with our accumulated knowledge, we foster a deeper understanding of the world around us. We begin to see connections that were previously hidden, identify patterns, and generate new ideas. This fertile ground is where moonshot moments take root.

For many centuries, numerous writers, philosophers, and polymaths around the world carried commonplace books to gather notes. When inspiration struck, they weren't scrambling to reconstruct the foundation of their knowledge. Instead, they had a rich and readily accessible oasis of information at their fingertips, allowing them to channel their creative energy into crafting a groundbreaking work. Modern, digital PKM is just an evolution of that superpower.

The promise of PKM lies in its ability to empower everyday individuals—not just tech wizards or academics—to become active participants in the knowledge economy. It equips us to seize those fleeting moments of inspiration and sparks of brilliance that can change the course of our lives and the world around us. As this chapter unfolds, we'll explore the practical aspects of building your personal knowledge oasis, exploring various PKM techniques and showcasing real-world examples of how individuals have harnessed its power to achieve remarkable things.

Gardens of Innovation: PKM in Action

One of the most powerful aspects of PKM is its ability to facilitate the cross-pollination of ideas. As a way of organizing notes and ideas, PKM can apply to any field, any background, or any project—everyone can benefit from PKM in their lives. By drawing from a diverse knowledge base, individuals can generate truly groundbreaking work and improve their level of accomplishment.

Sounds great, but what does that really mean? Let me give you some examples to show just how widely applicable a good PKM mindset can be.

Imagine an architectural firm tasked with designing a sustainable office building. Through their PKM system, they have access to research papers on passive heating and cooling techniques (grabbed from a recent engineering conference), some notes on historical case studies of beehive structures renowned for their natural ventilation (gleaned from an architectural history book), and even summaries of blog posts on biomimicry in design (discovered through online research). They even have a few notes they've dictated while watching a nature documentary last weekend with their children. By cross-referencing this information, the team designs a sustainable office building that utilizes passive techniques and natural ventilation, minimizing environmental impact and operational costs.

Perhaps a marketing manager for a new music streaming service is brainstorming user engagement strategies. Her PKM system holds a trove of relevant information: competitor analysis reports (gathered from internal documents), research on user behavior in social media platforms (sourced from marketing journals), and even insights on the psychology of music preference (culled from a fascinating podcast episode). By weaving together these diverse threads, the marketing manager develops a user engagement strategy for the music streaming service that combines targeted social media marketing with personalized recommendations based on user psychology and preferences, ultimately increasing user loyalty and retention.

Struggling to create nutritious and engaging meals for her picky eater, a stay-at-home-mom utilizes PKM. Her system includes recipes from cookbooks (categorized by dietary needs and allergies), a few summaries she scribbled down about blog posts on child development and healthy eating habits (gleaned from online research), and even voice notes on her child's favorite flavors and textures (gathered from the last tantrum he threw at the seafood restaurant). By cross-referencing this information, this mealtime magician crafts delicious and nutritious meals that cater to her child's preferences, promoting healthy eating habits from a young age.

Aiming to improve patient care through personalized medicine, a doctor utilizes PKM for his own personal improvement. His system might include medical research papers (categorized by specific conditions), patient medical history observations, and even notes on emerging treatment options gleaned from conferences. By analyzing this data and drawing connections between seemingly unrelated information, the doctor identifies potential risk factors for individual patients and tailors his treatment plans to maximize their effectiveness.

Facing a competitive job market, a young GenZ job seeker leverages PKM to stand out. His system holds company research notes (gathered from attending industry events), interview tips from career websites, and even personal branding strategies gleaned from online tutorials. This knowledge base allows him to tailor his resume and cover letter to each specific company, highlighting relevant skills and experiences while confidently navigating the interview process.

Snagged on a technical hurdle in developing a new product for one of the startups I help incubate, this entrepreneur uses PKM for guidance. My system contains notes on relevant engineering principles (learned from talks with a CTO of one of my other companies), case studies of similar product development challenges, and even online forum discussions with other entrepreneurs. By leveraging this collective knowledge base, I identified proven problem-solving strategies, adapted existing solutions to my specific needs, and overcame technical hurdles efficiently.

Yes, you probably noticed I changed perspective there! And that's because I'm just another case example: embracing a PKM mindset and system has revolutionized the speed and creativity I can bring to all the companies I work with. Like all these case examples, I have found the ability to achieve greater cross-pollination of ideas within myself and on my team by embracing a simple but highly effective organizational strategy for my time and my information.

From Commonplace Books to Digital Tools

Centuries before the digital age transformed how we access information, a powerful tool existed for cultivating and harnessing knowledge: the commonplace book.[2] These were not mere diaries or random collections of thoughts. Instead, meticulously crafted commonplace books served as a scholar's trusted companion, a curated archive of the mind's encounters with the world. Leather-bound volumes bulged with handwritten treasures: insightful excerpts from groundbreaking books, observations gleaned from daily life, and even artistic sketches sparked by a flash of inspiration.[3]

Writers like Virginia Woolf were avid commonplace book keepers. Woolf's notebooks overflowed with excerpts from the books she devoured, alongside her own reflections and observations on the world around her. These rich collections served as a constant wellspring of inspiration, fueling her literary genius.[4] Johann Wolfgang von Goethe, the literary giant behind Faust, employed a similar method. His commonplace books, two now resting at Harvard,[5] weren't a haphazard jumble of thoughts; they were meticulously categorized by topic, ensuring he could swiftly revisit a specific quote or observation that sparked his creative fire. Even the visionary science fiction author Octavia Butler employed this practice. For Butler, the commonplace book wasn't just about capturing grand ideas; it was a space where she could meticulously weave together threads of history, science, and social commentary into the groundbreaking narratives that would captivate readers for generations.[6]

Commonplace books transcended the literary realm. Polymaths like Leonardo da Vinci, the quintessential Renaissance Man, used them not just for artistic inspiration but also for scientific observations and engineering concepts.[7] These leather-bound volumes brimmed with sketches of fantastical flying machines nestled beside meticulously labeled anatomical drawings and detailed notes on the properties of light and shadow. Sir Francis Bacon, the philosopher and statesman, not only valued commonplace books, but actively championed a systematic approach to knowledge capture.[8] In his groundbreaking work, *Novum Organum*, Bacon laid the foundation for the scientific method. He emphasized the importance of meticulously recording and organizing observations to fuel discovery—turning the commonplace book into the concept of laboratory notes.

For Ben Franklin, the epitome of American ingenuity and one of my favorite moonshot moment makers, the commonplace book mirrored his boundless curiosity.[9] It wasn't just a single volume, but a collection of notebooks that served as a testament to his ever-expanding intellectual universe. Meticulously categorized entries showcased the depth of his exploration. Political treatises shared space with philosophical musings, scientific experiments meticulously documented alongside practical inventions like the lightning rod, a testament to his desire to bridge the gap between theory and application. Between the lines could be found pithy sayings, verses, and even jokes.[10] Franklin's commonplace book wasn't just a passive repository; it was a dynamic conversation with himself. Marginal notes sparked new ideas; connections drawn between seemingly disparate topics hinted at future inventions. This rich tapestry of knowledge became the fertile ground where his brilliance blossomed, a testament to the power of harnessing and nurturing intellectual curiosity.

One of the most fascinating examples comes from the renowned philosopher John Locke. A champion of reason and enlightenment, Locke believed that a well-maintained commonplace book was a cornerstone of intellectual development. In his detailed guide, *A New Method of Making Common-Place-Books*[11] (published in 1706), Locke outlined a system for organizing information by topic, with subcategories for specific details and references. He essentially wrote the book on commonplace books! His own volumes contained quotes from prominent thinkers, meticulously categorized by subject with detailed references for further exploration. Locke even included sections for recording observations from daily life and arguments encountered in his studies. This meticulously curated knowledge base served as a springboard for his seminal work, *An Essay Concerning Human Understanding*, a cornerstone of modern philosophy.

The rise of the digital age has ushered in a revolution in knowledge management. No longer confined to handwritten entries and leather-bound notebooks, our commonplace books have evolved into sophisticated digital tools. Instead of ink stained pages tinged with candle smoke and fingerprints, our digital applications allow us to capture information with ease, from web clippings and research papers to voice recordings and mind maps. These tools offer powerful capabilities for organization, tagging, and cross-referencing, transforming our knowledge into readily accessible databases.

PKM is also essential in helping to identify our values and goals, and then organizing our knowledge to help us work toward those goals—in that sense, PKM is a process of weighing value and making choices. Clinical psychologist Orion Taraban explains that

> Our mind is producing and reproducing on a moment-to-moment basis with respect to things in our environment. Value pertains to our personally relevant goals and where those goals are located in a nested hierarchy of goals. Because the human experience is one of playing many nested games simultaneously, something is of value when it is perceived to be more instrumental in achieving a personally relevant goal that is higher in the nested hierarchy of simultaneous goals. All of this perception of value is changing moment to moment and is mediated by memory, imagination, perception. It's a very complex, cognitive act. But it eventually creates a value coefficient: the sum of uncountable and instantaneous parallel calculation with respect to valuation. The issue is that nobody is aware of that process occurring. It occurs so deep beneath the threshold of awareness that most people are only dimly aware of one or two components of their decision-making process at most. That's usually for the best because more awareness can be overwhelming; we have a limited bandwidth especially with respect to our working memory and what we can attend to at any one time.[12]

In this way, PKM goes beyond mere digital storage and becomes an act of identity creation and value analysis. The true power of PKM lies in the act of curation and reflection and in helping us work through the decision-making process of determining what is valuable to us. Actively engaging with our accumulated knowledge—making connections, identifying patterns, projecting value, and drawing insights—cultivates a fertile ground for innovation. This is where moonshot moments, those transformative sparks of inspiration, have the potential to take root. Just as the cross-pollination in a well-tended garden leads to unexpected and vibrant blooms, actively cultivating our knowledge oasis fosters connections between seemingly disparate ideas, paving the way for groundbreaking discoveries.

The Foundations of PKM: The PARA and Zettelkasten Methods

With various PKM approaches available, how do you choose the one that unlocks your moonshot potential?

The key lies in understanding your own workflow and priorities. Some methods prioritize project-based organization while others focus on fostering long-term knowledge accumulation. I'm going to walk through some of the most effective and practical PKM mindsets, each of which approaches knowledge in a different way, but each of which also has great potential to inspire the kind of free-association thinking that is a core part of making moonshot moments happen. Let's first look at some of the broadest PKM approaches, which anyone can manage in any number of digital applications: Tiago Forte's PARA method and Sönke Ahrens' Zettelkasten method.

Tiago Forte's PARA method is one of the granddaddies of this field, offering a clear and action-oriented approach to PKM. "PARA" is an acronym (acronyms are quite popular in this field, as you'll soon see) and stands for "Projects, Areas of Interest, Resources, and Archive."[13] This simple framework categorizes information based on its function and lifespan, making it great for operationalizing all the notes and ideas you gather.

- **Projects:** These are specific endeavors with a defined start and end date. Your PKM system becomes a command center for project management, housing to-do lists, brainstorming notes, and relevant research materials.
- **Areas of Interest:** These are ongoing pursuits that don't have a clear endpoint. Perhaps you're passionate about sustainable living or fascinated by

the history of artificial intelligence. Your PKM system becomes a repository for articles, podcasts, and notes related to these long-term interests, allowing you to cultivate a deep understanding over time.

- **Resources:** This category holds reference materials with long-term value, like evergreen articles on productivity hacks or cheat sheets for specific software programs. Think of it as your digital bookshelf, readily accessible for future reference.

- **Archive:** Not everything needs to be actively managed. The Archive holds information that may not be immediately relevant but could be useful in the future. Old project materials, outdated research, or personal receipts can all be stored here, easily retrievable if needed.

The PARA method's appeal lies in its user-friendliness. With its clear-cut categories, even PKM newcomers can quickly establish a system for capturing and organizing information. This structure translates to action: the project focus keeps you laser sharp, ensuring your knowledge base directly fuels progress toward your goals.[14] Furthermore, PARA boasts flexibility. The categories can adapt and expand based on your specific workflow, making it a customizable system that grows alongside your needs. However, the PARA method isn't without its limitations. I find that its emphasis on projects prioritizes a linear approach, potentially stifling the creation of unexpected connections between seemingly unrelated ideas. Additionally, I don't find the project-centric focus ideal for those seeking to cultivate a broad and interconnected knowledge base across diverse areas.

The PARA method shines for individuals who thrive on structure and action. It's ideal for project managers, entrepreneurs, and anyone who wants a clear system for managing ongoing tasks and information relevant to specific goals. I have found that many small business owners, especially those working in creative industries, really benefit from the project-oriented structure.

In contrast, the Zettelkasten method, championed by Sönke Ahrens and inspired by the note-taking system of sociologist Niklas Luhmann,[15] takes a more open-ended approach to PKM.[16] At its core lies the idea of a *zettelkasten* ("notebox" in German) of diverse atomic notes: small, self-contained pieces of information captured on index cards or digital equivalents.[17] These notes are then linked together thematically, forming a dynamic web of knowledge.

The Zettelkasten method thrives on the notion of capturing fleeting ideas before they vanish. Zettelkasten is centered on the process of jotting down insights on individual note cards, each one a self-contained burst of inspiration. Here, focus reigns supreme—each card tackles a single concept, ensuring clarity and avoiding the pitfalls of information overload. But the beauty of Zettelkasten lies not just in capture but in connection. Each note receives a unique identifier, a fingerprint in your knowledge base. This allows you to weave a web of connections on the back of each card, using a simple notation system to link to related ideas. Over time, your knowledge base transcends static categories; it becomes a dynamic and ever-evolving organism, where connections emerge organically as you explore the rich tapestry of your own thoughts.

Unlike the PARA method's focus on project-based organization, Zettelkasten thrives on serendipity, fostering a landscape where groundbreaking ideas can blossom from the unexpected intersections of your accumulated knowledge.

The Zettelkasten method's true power lies in its ability to unlock the potential of nonlinear thinking. By capturing ideas as atomic units and then linking them based on thematic connections, you cultivate a fertile ground for serendipity. You are building a web of interconnected knowledge, where insights from seemingly disparate disciplines can spark groundbreaking discoveries. Unlike the PARA method's focus on project-specific information, Zettelkasten encourages the cross-pollination of ideas, fostering a creative environment where the next moonshot moment might just be a link away.

Furthermore, Zettelkasten excels at building a comprehensive knowledge base over time. As you accumulate atomic notes and weave connections between them, your understanding of a topic deepens and expands organically. This ever-evolving knowledge ecosystem becomes an invaluable resource, constantly enriching your perspective and fueling your intellectual growth.

However, the Zettelkasten method does come with a steeper learning curve. Mastering the art of atomic note-taking, where each idea is concise and self-contained, takes practice. Likewise, developing an effective linking system to foster meaningful connections between your notes requires dedication and experimentation. There's more of a time commitment on a daily basis, but that's because you are feeding a much larger garden.

The Zettelkasten method is ideal for those who crave a deeper understanding of complex subjects and relish the thrill of unexpected connections. It empowers researchers, writers, and anyone seeking to build a rich, interconnected knowledge base that fuels long-term learning and sparks groundbreaking ideas.

PARA and Zettelkasten, despite their differing organizational structures, share a strong belief in the power of capturing every insightful thought or piece of information. Both methods prioritize well-summarized notes that go beyond the immediate context. PARA might have you jot down a concise takeaway from a conversation in your "Projects" section, even if its full relevance isn't immediately clear. Likewise, Zettelkasten encourages condensing an interesting fact gleaned from a news article into a note, trusting that its future value will be revealed through connections with other ideas. Both methods are founded on the core belief that the act of summarizing and contemplating information during note-taking is crucial in allowing you to extract the essence of an idea and make it more readily connectable to future learnings, actually training your mind to be more open to associational thinking and creativity. This philosophy prioritizes the power of "potential"—transforming your knowledge base into a fertile ground where seemingly disparate concepts can blossom into unexpected insights and moonshot moments.

While PARA and the Zettelkasten method offer powerful approaches to knowledge management, I've also developed my own PKM system. This system, called Missions, Anchors, Vaults, and Identity (MAVI), takes inspiration from the PARA method but adds a crucial element—Identity. I find the MAVI system is best oriented toward those navigating career changes or working toward personal growth, those looking for structure as they question their life trajectory. MAVI's Identity section acts as a guiding light during this transformation. By crystallizing core values, long-term goals, and sources of inspiration, it empowers the user to make decisions aligned with their desired future self. This clarity ensures their Missions and Anchors actively propel them toward their aspirations while their Identity provides mantras and inspirational sources serve as a wellspring of motivation when faced with challenges.

In essence, MAVI's Identity section isn't just about self-reflection; it's about self-creation. It empowers you to design the future version of yourself and ensures every element within your PKM system actively contributes to that transformation. This laser focus on core values and long-term goals makes MAVI an invaluable tool for anyone navigating the exciting, yet often uncertain, path of personal growth. MAVI recognizes that significant life transitions often involve reevaluating core values and aligning actions with a desired future self. Here's how each component of MAVI fosters this focus on personal growth:

- **Missions:** Similar to PARA's Projects, MAVI's Missions encompass the projects, tasks, and habits that propel you toward your goals. However, MAVI encourages you to constantly assess whether these Missions align with your evolving Identity.

- **Anchors:** These represent the foundational elements that shape who you are, such as responsibilities, commitments, and important relationships. Unlike PARA's Areas, MAVI's Anchors are not static. They can be reevaluated and adjusted as you navigate your personal growth journey.

- **Vaults:** Just like PARA's Resources, MAVI's Vaults serve as interconnected databases to store the knowledge you need to succeed.

- **Identity:** This is the cornerstone of MAVI, absent from the PARA method. Your Identity encompasses your north star—your core values, long-term goals, motivators, mantras, and sources of inspiration. By clearly defining your Identity within MAVI, you ensure every Mission, Anchor, and Vault you create fuels your journey toward becoming the best version of yourself.

Tiago Forte argued that those who fall in love with PKM ultimately become proselytizers themselves, eager to share what they've learned.[18] This has been my experience, and I can only hope that my success in developing an identity-oriented PKM system will help others, from GenZ entrepreneurs to ready-to-retire Boomers, get the most out of the ideas, notes, and goals they manage within their MAVI systems.

PARA, MAVI, and Zettelkasten are all gloriously adaptable. Their core principles can be implemented using a surprising variety of tools. Physical note cards meticulously organized in boxes, or even colorful sticky notes plastered on a wall, can be an effective way to get started. However, the true power of these systems shines brightest in the digital realm. Digital tools allow for effortless expansion of your knowledge base. Notes can be effortlessly linked, searched, and reorganized, fostering a dynamic interplay of ideas. There's a wealth of fantastic apps available to supercharge your PARA or MAVI workflow. Additionally, there's a whole ecosystem of apps that integrate seamlessly into your PKM system, allowing you to effortlessly clip interesting articles for later reference, manage tasks derived from your notes, and more. I'll talk about how to navigate this cornucopia of PKM options in a bit. But, as another option, there's an exceptional pre-built PKM system that may be well suited for shepherding your moonshot moments: August Bradley's Pillars, Pipelines & Vaults (PPV) for Notion.

If You've Got the Notion: Bradley's PPV

Some PKM enthusiasts have taken things a step further, crafting specific implementations tailored to popular software and sharing them with others. This removes the initial hurdle of building your own organizational structure within an app, and allows for more complex filing systems (since you don't have to build them yourself).

The beauty of this approach lies in its simplicity. Instead of spending hours customizing an app to fit your needs, you can jump right in and start capturing and connecting your ideas. The best example is August Bradley's PPV system designed for Notion. Bradley's system offers predefined workflows within their respective platforms, essentially upgrading on the PARA system by bringing in the best elements of Zettelkasten.

Pillars, Pipelines & Vaults (PPV) is a targeted PKM system, designed by successful business leader and Notion aficionado August Bradley to transform a

PKM setup into a powerhouse for focus, alignment, and knowledge retrieval, propelling users toward those ambitious moonshot goals.[19] Bradley's framework revamps the PARA approach into an easy-to-use system of information flows and storage spaces. He has pre-built his approach into a streamlined Notion system, doing away with the clutter of multiple note-taking products. Bradley's PPV breaks down into three parts:[20]

- **Pillars:** Pillars are broad categories of your day-to-day life, like "growth," "home life," and "business." PPV's pillars act as your categorical compass, guiding you through life's daily complexities. Every piece of information, every task, every goal you capture falls under one of these pillars, creating a clear organizational structure. No more wasting time searching through a disorganized knowledge base—you know exactly where to find the information you need to fuel your next moonshot.

- **Pipelines:** While PKM excels at knowledge management, it doesn't inherently translate that knowledge into action. This is where pipelines step in. Pipelines represent the actionable side of the PPV system. Think of them as the assembly line for your project. Pipelines house your to-do lists, project plans, and goals, all categorized by your pillars. This ensures your daily actions directly contribute to your overarching moonshot aspirations. No more spinning your wheels on tasks that don't move the needle. Pipelines keep you llaser focused on the actions that truly matter, transforming your knowledge into a springboard for innovation.

- **Vaults:** Vaults function as your personal knowledge archive, meticulously storing articles, ideas, and insights you've captured from various sources. But vaults go beyond mere storage—they're designed for intelligent retrieval. PPV's vaults ensure these connections are readily available when inspiration strikes. PPV employs a system that surfaces relevant information precisely when you need it most.

Let's walk ourselves through the PPV system in action. As you meticulously build your PKM system, capturing every relevant piece of information for your moonshot project, the pillars ensure a clear organizational structure. You then translate this knowledge into actionable steps using pipelines, creating a road map of tasks and goals aligned with your moonshot vision. Finally, the vaults become your ever-evolving knowledge bank, constantly enriching your understanding and sparking unexpected connections that propel your moonshot idea forward.

PPV isn't just about organization or productivity. It's about harnessing the power of knowledge to fuel innovation while connecting to your overarching values and core categories at all times: never losing sight of the pillars that uphold both your daily life and long-term goals.

The PPV system boasts a strong focus on goal alignment, but it's not without weaknesses. PPV relies heavily on using a specific tool like Notion. While powerful, Notion has a learning curve. PPV can feel overwhelming at first, with its three main components (pillars, pipelines, and vaults) and the additional structure it imposes on your existing PKM system. Implementing and maintaining this system can require a significant time investment, especially for those already comfortable with their current workflow. Additionally, the system can lead to information overload within the vaults if not managed properly. Ultimately, the effectiveness of PPV depends on individual needs and preferences. Consider adapting the system to best suit your workflow and prioritize ease of use over rigid adherence to every element. Remember, the goal is to leverage knowledge for moonshot success, not become a prisoner to complexity.

While PPV offers excellent predefined structures built within specific applications (Notion specifically), this approach is definitely less flexible than completely open-ended PKM systems. You're trading flexibility for a well-oiled machine, meticulously designed to maximize efficiency and achievement. With PPV, you'll gain a system battle-tested to propel you toward more accomplished goals and those coveted moonshot moments. Remember, regardless of the system, success ultimately hinges on understanding the note-taking and organization tools that best suit your individual workflow within your chosen PKM approach.

Tools for Your Inner Polymath

Between these four approaches—the PARA, Zettelkasten, MAVI, and PPV systems—anyone can find a PKM mindset that fits their needs and goals for self-improvement. The journey to a powerful PKM system starts with a clear vision. First, identify your goals.[21] What knowledge do you crave? Are you aiming to master a new skill, ignite innovation in a specific field, or simply become a lifelong learner? Moonshot moments often thrive in fertile ground. Keep them in mind as you define your desired outcomes.

The cornerstone of a successful PKM journey lies in selecting the perfect tool to house that intellectual ecosystem. This decision can feel overwhelming, given the vast array of options available. To navigate this landscape effectively, it's crucial to prioritize introspection and understand how your specific needs and workflow intersect with the functionalities offered by various PKM tools. There are many worthwhile questions to ask yourself along the way.

Start by taking stock of the features that resonate most with you. Are you a meticulous note-taker who thrives on robust tagging and linking capabilities, allowing you to effortlessly connect ideas across your knowledge base? Or perhaps spaced repetition, a technique for enhancing long-term memory, is a key element in your learning process. Take a moment to look within and consider your organizational preferences. Does a structured, hierarchical approach with clear categories and subcategories bring you joy? Alternatively, do you find inspiration in the free-flowing, nonlinear world of thought exploration, where connections emerge organically?

If we turn back to Tiago Forte, the father of PARA, he offers a few suggested features to look for in software for your PKM system. First, he suggests choosing a system that allows for multimedia note-taking. After all, you never know when you'll want to add a quick sketch or a brief audio note. He also suggests choosing a software that allows for open-ended note-taking, as opposed to slide decks or spreadsheets with enclosed columns and cells. Finally, Forte reminds us that our notes will be messy, and they don't need to be precise. So it's important to prioritize apps that allow for informality and quick idea capture rather than to focus on strict templates.[22]

These points are why I always consider a tool's accessibility when I try out new PKM and note-taking programs. Does it seamlessly integrate with my workflow across various devices? Is it readily available on my desktop, mobile phone, and web browser, ensuring smooth information capture and retrieval regardless of location or device? And of course, cost is another important factor. Freemium models offer a good starting point, while more feature-rich options might require subscriptions. You should always test out the free version of a tool to see if it matches your personal style of note organization.

Remember, the ideal PKM tool should feel like an extension of yourself, not a foreign entity disrupting your established habits. If PKM doesn't fit naturally into your everyday life, you won't use it constructively. So seek a PKM program that integrates effortlessly into your workflow or daily life, feels intuitive to use, and caters to your specific needs and preferences. By carefully considering these factors, you'll be well on your way to selecting the perfect PKM tool and embarking on a rewarding journey of knowledge management.

While I have worked my way through many PKM system technologies—other people play video games, I play around with new PKM systems!—there is one popular PKM option that I really appreciate, which has had real staying power in the PKM scene: Notion. Notion is also used in business settings but has so much to offer an individual user from any walk of life. Notion stands out as a versatile workspace app. Its ability to morph into a PKM powerhouse is undeniable. Databases allow you to meticulously categorize your notes while wikis provide a space for in-depth exploration of complex topics. Project management tools keep your workflow organized, ensuring your ideas translate into actionable steps. If you're drawn to August Bradley's PPV system, Notion's flexibility provides the perfect canvas to bring his Pillars, Pipelines & Vaults to life. But I'd recommend Notion for any PKM system, from MAVI and PARA to whatever other system you develop personally.

Some users find Notion too complex, or too much like a wiki-entry, and prefer other options. For example, ClickUp is a project management powerhouse that can be equally adept at managing your knowledge. ClickUp's strength lies in its highly customizable lists and views. This allows you to tailor your PKM system to your specific needs, ensuring a perfect fit for your workflow. ClickUp also boasts robust integrations with a variety of other tools, enabling you to connect your knowledge base to external resources and streamline your information gathering process.

Delving deeper into the PKM app landscape, several options cater to specific preferences. If you value privacy and offline access, Obsidian shines with its local-first architecture, markdown editing, and robust backlinking system. This fosters a knowledge base that is completely under your control and perfect for meticulous organization.[23] Logseq falls into a similar category, offering a graph-based interface ideal for visually connecting ideas. This free, open-source app also boasts built-in spaced repetition, a powerful tool for retaining information over time.[24]

For those prioritizing nonlinear thought exploration, Roam Research stands out. Its unique interconnected note structure allows you to build a web of ideas without a predefined hierarchy. While it requires a subscription, Roam Research empowers researchers, writers, and anyone who thrives on brainstorming and making unexpected connections.[25]

Following the path of scholars and geniuses of the past, Zotero caters specifically to the needs of academics and researchers. This reference management software excels at organizing and citing research materials, making it an invaluable tool for scholarly pursuits.[26] For those already comfortable with established names, Evernote remains a solid choice with its comprehensive feature set, including web clipping and collaboration tools. Bear offers a beautiful and user-friendly alternative, ideal for those who prioritize aesthetics and simplicity in their note-taking experience.[27] RemNote integrates seamlessly with any PKM system, providing science-backed spaced repetition to supercharge your knowledge retention.[28]

Overwhelming as it can be, I'm quite glad there are so many options already, with more PKM applications coming out each year: this demonstrates humanity's clear desire to upgrade ourselves, as individuals, to be more engaged with our ideas and more open to freer thinking. PKM systems are the first link between the more hyper-cooperative societies of the future and our own personal growth within them. This is your opportunity to join in that process and embrace your path to moonshot moments. Once you've chosen your PKM companion, it's time to start collecting! Capture everything that sparks your curiosity—notes, ideas, articles—and organize them within your system using the available tools. Remember, consistency is key. The more you feed your PKM system, the more it empowers your learning journey.

Become a PKM Rockstar

I'd be remiss not to share some of the battle-tested PKM tricks I've accumulated on my knowledge management journey. These advanced techniques aren't just about bells and whistles—they're powerful tools specifically designed to maximize your productivity while simultaneously reducing the feeling of being overwhelmed by information overload.

Here's why: by their very nature, basic PKM practices like note-taking and organization help you capture and categorize information. But these advanced techniques take things a step further. Active note-taking with linking and elaboration, for example, allows you to not just capture information but to truly understand and connect it to existing knowledge. This fosters deeper comprehension, reducing the need to revisit the same source material repeatedly. Advanced PKM techniques will empower you to work smarter, not harder. You'll spend less time sifting through information and more time leveraging it to achieve your goals. By incorporating these advanced techniques, you can transform your PKM system from a passive archive into a dynamic springboard for moonshot moments, all while minimizing the feeling of information overload and maximizing your overall productivity.

The first tip is to move beyond passive note-taking, a mere transcription of information, and instead actively engage with the knowledge you capture. When reading an article on, say, sustainable architecture, don't just copy quotes. Jot down your own thoughts, questions, and potential applications. Ask yourself, "How can I use this information?" Link these notes to related ideas in your system, perhaps to an earlier note on urban planning or a brainstorming session on eco-friendly housing solutions. This cross-pollination fosters a web of interconnected ideas within your PKM system. By actively engaging with your notes and fostering these connections, you transform your PKM system from a static archive into a dynamic launchpad for groundbreaking ideas.

Moonshot moments rarely emerge fully formed. Unleash your inner genius with visual tools like mind maps. Say you're a writer with a moonshot goal of crafting a captivating science fiction novel. Start a mind map with "Sci-Fi Novel" at the center. Branch out with ideas for genres, settings, characters, and plot twists. Visually connecting these concepts can spark unexpected combinations and ignite your creativity. Even a simple sketch of a fantastical creature or a futuristic cityscape can become the catalyst for your next literary masterpiece. When you make mind maps, drop them into your files in order to remind you of overarching themes, values, and visions later down the road.

The human brain is notorious for forgetting. Don't let your PKM system become a graveyard of forgotten ideas. Integrate spaced repetition, a scientifically proven technique for knowledge retention that utilizes optimal timing of note reviews to train your brain and maximize retaining information.[29] Some tools like Anki or RemNote even allow you to schedule reviews for your most important notes, ensuring critical information relevant to your moonshot goals stays top-of-mind. Imagine a software developer meticulously reviewing notes on cutting-edge algorithms through spaced repetition. Weeks later, while grappling with a complex coding challenge, a seemingly unrelated note on a novel data structure sparks a flash of inspiration, leading to a breakthrough solution. By leveraging spaced repetition, you ensure the vital knowledge within your PKM system remains readily accessible, priming you for those moments of unexpected brilliance that can turn your moonshot goals into reality.

Don't wait for the lightning bolt of inspiration to strike solely within the confines of your PKM system: embrace spontaneity. The real world is brimming with potential sparks of brilliance. Set up external triggers to capture these fleeting ideas on the fly. Keep a note card tucked in your wallet and a voice recorder app readily available on your phone. A quick jot-down during your commute or a voice memo capturing a snippet of overheard conversation on your lunch break could be the seed for your next breakthrough. What about a musician out for a walk, noticing the unique call of a particular bird? With a practiced flick of the wrist, they whip out their phone and record the melody. Back at their workstation, that recording transforms into the captivating hook for their next hit song. By incorporating these external triggers, you will weave a richer tapestry of inspiration into your PKM system, priming yourself for unexpected moments of creative genius.

Information overload is a real threat in today's digital age. Not all content deserves a coveted spot in your meticulously crafted PKM system, so be savvy in your media consumption and note-taking. Be mindful of the sources you consume. Transform yourself into a discerning connoisseur of information, actively curating a list of high-quality resources, credible websites, and thought leaders in your specific field. This ensures the foundation of your knowledge base is strong, reliable, and trustworthy. Think of a data analyst who prioritizes populating their PKM system with research papers from esteemed academic journals and meticulously avoids the alluring clickbait articles and social media "hot takes" flooding the Internet. By cultivating a discerning approach to information consumption, you foster a PKM system built on a bedrock of credible data, empowering you to make sound decisions and fuel groundbreaking ideas.

Complacency is the enemy of progress. Don't let your PKM system become a stagnant repository of information. Schedule regular reviews to assess its effectiveness. Forte suggests weekly and monthly reviews, plus reviews at the beginning and end of new projects.[30] Are your current organizational methods still serving you well? Are there any outdated notes or irrelevant information cluttering your system, hindering your ability to find what you need? As a writer, I dedicate a specific time each week to revisiting my PKM system. With a critical eye, I prune unnecessary notes, identify areas for improvement in my organizational structure, and actively seek out new ways to leverage my PKM system's capabilities. This commitment to constant evolution ensures my PKM system remains a powerful tool for growth, not a dusty archive of forgotten ideas. By embracing regular reviews, anyone can position themself to reap the long-term benefits of their PKM efforts and transform it into a dynamic springboard for lifelong learning and innovation.

By incorporating these advanced techniques, you can transform your PKM system from a static archive into a dynamic springboard for moonshot moments. Active note-taking fosters deeper engagement with information, while mind mapping and visualization unleash creative potential. Spaced repetition ensures crucial knowledge stays top-of-mind, and building a knowledge graph allows you to uncover hidden connections and spark groundbreaking ideas. Remember, consistency is key. The more you feed your PKM system with these advanced techniques, the more it empowers you to make those unexpected leaps of brilliance and turn your moonshot goals into reality.

Other PKM Cheat Codes: Time Blocking and Gamification

Personal knowledge management is a powerful tool for conquering information overload, streamlining workflow, and fostering a culture of exploration—all crucial ingredients for the moonshot mindset. But, just like in any game, even the most well-organized player can benefit from a few power-ups. Here, I'd like to show you two such productivity boosters that amplify the efficacy of my PKM: time blocking and gamification. Think of them as upgrading your character so you can play the game of moonshot innovation even better.

Time blocking is a productivity technique where you schedule specific blocks of time in your calendar for dedicated tasks. It's like carving out mini arenas within your day, each designed for a specific kind of activity.[31] It's easy to envision a writer blocking out a morning session for focused writing, followed by an afternoon block for research and interviews; or an entrepreneur, blocking off an evening each week just to read about emerging trends in global markets. This approach eliminates the constant context switching that drains mental energy and allows for deep dives into specific tasks.

One of the key advantages of time blocking is its ability to combat procrastination.[32] By pre-allocating time for specific tasks, you remove the decision-making burden in the moment. No more staring at a to-do list, unsure where to begin. Time blocking also fosters a sense of control and ownership over your schedule, reducing anxiety and boosting motivation. However, time blocking isn't without its limitations. It requires a certain level of discipline and self-awareness. Underestimating how long a task will take can throw off your entire schedule. Additionally, the unexpected always finds a way in, demanding flexibility. The key is to be realistic when allocating time blocks and remain open to adjustments when necessary.

The connection between PKM and time blocking is undeniable. A well-organized knowledge base, meticulously built through PKM, becomes the fuel for your time-blocked activities. PKM ensures you have the knowledge at your fingertips when you need it most, allowing you to maximize the productivity of each time block. It also becomes much easier to find time to review your notes, organize recent notes, and conduct monthly audits if you have a clear block set aside for maintaining your PKM garden.

Gamification involves incorporating gamelike elements—points, badges, leaderboards—into nongame contexts to increase engagement and motivation. Turn your to-do list into a quest, with completed tasks awarding points that unlock new levels or badges.[33] This playful approach can make even mundane tasks feel more engaging and rewarding. After all, I started this entire section talking about cheat codes to upgrade your character, and you probably felt a little tinge of excitement at the prospect. It's that energy, from our childhood when things felt easier and more carefree, that we want to add to spice up our PKM experiences, to make them fun and more productive.

Studies have shown that gamification can trigger the release of dopamine, a neurotransmitter associated with pleasure and motivation.[34, 35] The act of completing tasks and receiving rewards activates the reward system in the brain, creating a positive feedback loop that encourages you to keep going.

However, gamification can backfire if the reward system is poorly designed or feels artificial. Additionally, the focus on points and awards can sometimes overshadow the intrinsic value of the task itself. The key is to find a gamification approach that resonates with you and complements your existing PKM system.

A writer can leverage gamification by awarding points for completed chapters, giving herself rewards for milestones like "10,000 words written" or "First draft complete" and transforming the writing process into a gamified quest for their moonshot manuscript. This approach applies beyond creative fields. An entrepreneur can design a PKM system that awards points for completed sales calls or milestones achieved, gamifying the daily grind of running a startup. Similarly, a language learner can design a PKM system that tracks vocabulary learned or grammar exercises completed, awarding points and badges to gamify the process of language acquisition.

Even a fitness enthusiast can utilize PKM to track workouts, with points awarded for completed routines and badges for reaching fitness goals, making exercise feel more like a game and less like a chore.[36] Sound like your favorite fitness tracker? That's because the devices you are already using to manage your personal data have embraced gamification in your PKM fitness systems. We'll talk a little later about what we can learn from the success of these early ventures into wearable PKM, but for now, just relish that most recent "steps badge" you earned!

Combining PKM with time blocking and gamification creates a powerful productivity symphony, each element amplifying the effectiveness of the others. PKM provides the organized knowledge base, time blocking ensures focused action within dedicated timeframes, and gamification injects a dose of motivation and fun. By harnessing the power of PKM, time blocking, and gamification, you can not only boost your productivity but also cultivate the fertile ground necessary for moonshot moments to flourish. So, equip yourself with these power-ups, embark on your innovation journey, and unleash the full potential of your mind.

Setting Yourself Up for Moonshot Thinking

A breakthrough idea sits on the horizon—a moonshot with the potential to transform your field. Excitement courses through you, but a cluttered mind chock-full of unorganized information threatens to stall your launch. Then your

PKM acts as a filter, a system for capturing, organizing, and readily accessing the information that truly matters to your moonshot pursuits.

But PKM goes beyond mere organization. It streamlines your workflow, ensuring you can act swiftly when inspiration strikes. Picture an entrepreneur with a revolutionary new app idea. They've captured the core concept in their PKM system, along with detailed notes on potential features, competitor analysis, and target user demographics. This readily accessible knowledge base allows them to move quickly from ideation to execution. They can develop a prototype, pitch to investors, and launch their moonshot venture with a laser focus, fueled by the efficient information management facilitated by PKM.

The link between PKM and productivity is undeniable. Studies have shown a strong correlation between productivity and happiness.[37] When we feel accomplished and in control of our tasks, our overall well-being flourishes. A happy mind is a fertile ground for moonshot ideas.[38] PKM empowers this positive cycle by giving you the tools to conquer your workload, fostering a sense of accomplishment that fuels your creative spirit and propels you toward even bigger goals.

The beauty of PKM lies in its compatibility with other productivity strategies. It seamlessly integrates with time blocking and time management techniques. Imagine a writer with a moonshot goal of finishing their magnum opus. They utilize PKM to organize their research materials, character profiles, and plot outlines. They then integrate this knowledge base with their time blocking schedule, allocating dedicated writing sessions throughout the week. PKM empowers them to maximize their writing windows by ensuring all the information they need is readily at hand. This isn't hard for me to imagine at all—this is the very process I had to work through to get this manuscript from my mind to the screen and off to your page!

But PKM isn't just about capturing information; it's about using it as a potent tool for exploration and experimentation, vital ingredients in the moonshot recipe. Scientists, inventors, and entrepreneurs all rely on a deep well of knowledge to fuel their moonshot pursuits. PKM provides a platform for them to not only store but also to actively engage with their knowledge. Advanced techniques like mind mapping and knowledge graphs allow them to visualize connections, identify patterns, and spark unexpected ideas. By visually linking diverse concepts, you can identify potential synergies and explore innovative solutions, potentially leading to the development of a groundbreaking new software application, or a new approach to teaching kids financial skills, or just the best recipe for apple pie ever.

Millions of individuals across various professions are harnessing the power of PKM to achieve remarkable things. Writers use PKM to organize complex storylines and characters, leading to the creation of captivating novels. Entrepreneurs leverage their PKM systems to develop innovative business models, disrupting industries and changing the world. Scientists meticulously manage their research using PKM, leading to groundbreaking discoveries and advancements in medicine, technology, and countless other fields. This is already happening at an individual level with such commonality that it is becoming a cultural movement with our society and species.

The impact of PKM extends far beyond individual achievement. By empowering individuals to manage, connect, and leverage their knowledge more effectively, PKM fosters a collective wellspring of creativity and innovation. I am hopeful that I have shown that PKM is more than just a fancy term for taking notes. It's a philosophy, a methodology, and a powerful tool for maximizing your productivity and propelling you toward achieving your moonshot goals.

As we embark on the next chapter of our exploration, I'll take you on a fascinating trip to the realm of psychedelics. It's easy to write these drugs off as forgotten relics of the 1960s, but I'll show you an emerging world of scientific breakthroughs in the battles against mental illness. These mind-altering substances also have the potential to unlock new avenues of thought, to shatter the confines of conventional thinking, and to foster a deeper, more associative approach to problem-solving. As we thrive with PKM today, could psychedelics be the next frontier in expanding our openness to ourselves, the world, and the possibilities around us, further increasing our capacity for moonshot thinking?

Chapter Seven

WE NEED TO CHANGE OUR MINDS: A CASE FOR PSYCHEDELICS

When Nobel Prize-winning chemist Kary Mullis died in 2019, his university publication ran a bold headline calling him an "Intolerable Genius" and arguing, "Kary Mullis revolutionized biology and pissed everyone off. Now that he's dead, how should we remember him?"[1] I don't know about you, but I'd like to hope my alma mater has a kinder opening to my obituary article—if I ever do become as important, and controversial, as Kary Mullis had become.

Before his groundbreaking discovery of polymerase chain reaction (PCR, a process that allows us to rapidly create millions of copies of a specific DNA segment for testing), Kary Mullis was a biochemist with an inquisitive mind and a penchant for independent thinking. He pursued chemistry in college, eventually earning a PhD from the University of California, Berkeley. There, Mullis gained a reputation as a strong user and supporter of psychedelics. In a Q&A interview Mullis once said, "Back in the 1960s and early 1970s I took plenty of LSD. A lot of people were doing that in Berkeley back then. And I found it to be a mind-opening experience. It was certainly much more important than any courses I ever took."[2] No, this probably isn't why his alma mater was so harsh (though I doubt it won Mullis many hearts and minds).

Following his academic pursuits, he worked for various research institutions, including a California biotechnology firm called Cetus. In fact, Mullis's friend and fellow researcher Tom White secured the job for Mullis at Cetus on his psychedelic credentials. "I said, if you hire this guy Mullis, he's an excellent synthetic chemist. I knew he was a good chemist because he'd been synthesizing hallucinogenic drugs at Berkeley."[3] Mullis was quickly unpopular at Cetus, however, because of his erratic behavior and generally aggressive attitudes: Mullis certainly was not the stereotype of the laid-back hippie.

It was during this period, in 1983, that Mullis found himself stuck on a problem: how to rapidly amplify specific segments of DNA. According to his Nobel Prize speech, he was driving up to his cabin in the woods (where he was well-known for indulging in psychedelics with his friends) when he had what he described as his "EUREKA!" moment.[4] Pulled off on the side of the road, writing on paper from his glove compartment, Kary Mullis solved one of synthetic biology's greatest problems. It was the very definition of a moonshot moment.

This sudden burst of clarity, fueled by both scientific inquiry and a nontraditional state of mind, led to the revolutionary invention of PCR, a technique that has since become fundamental in many fields, from medicine to forensics. This invention would forever change the landscape of molecular biology and ultimately earn Mullis the Nobel Prize in Chemistry in 1993.

Kary Mullis would go on to have an unusual "career." Considered a liability for sloppy lab techniques and a penchant for office lovers' drama, Mullis was encouraged to leave Cetus with a $10,000 bonus before he had even won his Nobel Prize. He was unemployed when he won the award; Cetus sold his PCR technique on the open market for a cool $3 million.[5] Mullis would go on to infamy, serving as a DNA expert at the OJ Simpson trial and making controversial statements about AIDS at international conferences at the height of the African AIDS crisis of the 1990s. Friends would abandon him, departments wouldn't hire him, and his few peer-reviewed papers are highly speculative.

Kary Mullis is considered by many to be a fluke, a one-hit-wonder scientific genius who lacked rigor or merit in a scientific community obsessed with publication stats and towing the peer-reviewed line. But Kary Mullis is a perfect example of the power of psychedelics to open our species up to moonshot moments: if even average scientific minds can be propelled to pure genius through the use of psychedelics, what would the world look like if we embraced them in scientific, clinical, and intentional ways?

The question of whether psychedelics can spark game-changing ideas is not a matter of speculation. Kary Mullis's insistence that his experiences with psychedelics allowed him to expand his mind enough to discover PCR serves as a powerful testament to their potential. Throughout this chapter, we will dive into other compelling examples where psychedelics played a direct role in birthing moonshot moments, forever altering the course of science and human understanding.

The Science Behind the Experience

Before we journey through the rich history of psychedelics, let's first define what we're actually talking about. Psychedelics are a class of substances that profoundly alter perception, mood, and cognitive processes. This often includes changes in sensory perception—think vibrant colors or swirling patterns—along with shifts in how one experiences the self and one's relationship with the world.

Among the most widely discussed psychedelics are psilocybin mushrooms, LSD (lysergic acid diethylamide), ayahuasca, 5-MeO-DMT, ibogaine, MDMA (commonly known as ecstasy), and ketamine. Each of these substances has distinct properties and applications, especially within medicinal and therapeutic contexts. For example, psilocybin mushrooms and LSD are classic psychedelics that primarily work on serotonin receptors, inducing powerful alterations in consciousness and perception. Ayahuasca, a traditional plant medicine from the Amazon, combines two key ingredients to produce a deeply introspective and often spiritual journey. 5-MeO-DMT, derived from the venom of the *Bufo alvarius* toad or synthetic sources, offers brief yet intense experiences often described as transcendent.[6] Ibogaine, sourced from the iboga shrub native to

Central Africa, is known for its potential in treating addiction, particularly to opioids. I'll explore ibogaine a bit more later as we discuss current therapeutic treatments offered to combat veterans.[7]

It's also important to distinguish between the terms *psychedelics* and *plant medicine*. While plant medicines like ayahuasca and psilocybin mushrooms originate from natural sources, ketamine, LSD, and MDMA are synthesized in laboratories and are not considered plant medicines. Still, laboratory-produced treatments offer the opportunity to treat several issues. MDMA, a psychedelic often categorized as an empathogen, enhances feelings of emotional connection and has shown significant promise in treating PTSD.[8] Ketamine, a dissociative anesthetic that technically does not fall under the classical definition of a psychedelic, is used in controlled doses to alleviate treatment-resistant depression and suicidal ideation. MDMA and ketamine are an essential part of the discussion of psychedelic therapy as they often serve as underground drugs used during treatment to prepare patients for more intense psychedelic therapy with true psychedelics.

By understanding this breadth, we can better appreciate the diversity within this field and the nuanced ways these substances are redefining modern mental health care and personal exploration. So that we don't get overwhelmed by all the options, I'll primarily focus on the two substances most commonly used in psychedelic therapy, as they also cause similar effects on the brain that are relevant for moonshot thinking: LSD and psilocybin mushrooms. But there's certainly exciting progress being made in the use of MDMA and ibogaine in therapeutic treatment, so we'll touch on those as well.

I want to take a moment to say that while the potential of psychedelics is incredibly fascinating to me as a writer and philosopher, I'm just here as a facilitator of information, not a medical professional. I am not a doctor, and I am not a medical expert! This book is not intended as an endorsement of drug use but rather as an exploration of ongoing scientific research. If you're curious about the potential of psychedelic therapy, remember to always prioritize your safety and well-being: consulting with a qualified healthcare professional is crucial to determine if that path is right for you. Professionals can provide guidance and ensure a safe, supported environment for exploration. I encourage everyone to have an open mind about psychedelic medicine and to embrace it if a professional prescribes it, but please do not confuse my commentary for medical expertise.

Also, to all young readers: please, don't do drugs. While we now have a very good idea what various drugs, including psychedelics, can do to a fully formed adult brain, we know very little about the impact of psychedelics on teenagers or children. With prolonged scientific research on the effects of psychedelics in therapy for children—yet another reason we desperately need more psychedelics research—we can have a clearer idea of the effects of psychedelics on brains in development. Until then, though, this text's discussion of psychedelics is always in the context of adult experiences, and again, always under the supervision of a medical professional.

Psychedelics obviously have a profound effect on the adult brain, but their impact is a combination of temporary physical changes and the profound perspective changes that come with the "trip." In other words, the physical effects of psychedelics are only meaningful in so far as they cause profound thought.[9] There is an overarching consensus across scientists who have explored psychedelics: the drugs have short duration effects (4–8 hours); they cause no permanent damage or even long-term physical effects; they are nontoxic and, perhaps most importantly, non-addictive. According to researcher Dr. David Nutt, "psychedelics overall are some of the safest drugs we know of . . . it's virtually impossible to die from an overdose of them; they cause no physical harm; and if anything, they're anti-addictive."[10] Many users require only one to three trips to have truly life-changing, deeply meaningful results but with no long-term effect on the body. That's because of the ways that psychedelics change our brain chemistry.

When we ingest psychedelics like psilocybin, ayahuasca, or LSD, they set off a fascinating chain reaction within the brain, particularly affecting two key networks: the Default Mode Network (DMN) and the Salience Network. Often referred to as our "self-referential network," the DMN is active when our minds are at rest, allowing us to engage in introspection, daydream, and construct our sense of self. We didn't even know the DMN existed until 2001; leading psychedelics expert Robin Carhart-Harris didn't even draw the first connections between psychedelics and the DMN until 2010.[11] The DMN essentially serves as the brain's internal chatterbox, constantly ruminating on the past, planning the future, and making judgments about ourselves and the world around us. The DMN controls our sense of self, maintaining the "who we are" at all times.[12]

Psychedelics like psilocybin, ayahuasca, and LSD disrupt the DMN's usual activity.[13, 14, 15] Studies suggest they decrease communication within the DMN, essentially quieting that internal chatter. This can lead to a feeling of ego dissolution, a loosening of our rigid self-perception, and a sense of interconnectedness with everything around us. Loosening the DMN's grip breaks down the barrier between the ego and other forms: people feel less separated from others, from the universe, and from nature,[16] leading many to experience different levels of profound connection and revelation.

Additionally, psychedelics affect what's called our "Salience Network." This network acts as a filter, prioritizing incoming information and directing our attention to what's most relevant at the moment. It helps us distinguish between a car horn and a friend's voice, for example. The Salience Network also affects our sense of our internal processes (like hunger or tiredness), our sense of empathy, and some of our emotional control.[17] Studies have shown that deep meditation specifically affects the Salience Network, explaining why Buddhist monks and others who practice extreme meditation also experience some of the same self-dissolution.[18, 19]

Psychedelics greatly increase activity in the Salience Network.[20, 21] This heightened sensitivity can lead to intensified sensory experiences, such as seeing vibrant colors in music or feeling a profound connection to nature. The increased salience may also explain why these substances can sometimes trigger emotional processing and bring suppressed memories to the surface. Perhaps most importantly, it helps us let go of existing prioritizations, allowing us to decide new priorities when considering new ideas.

Additionally, psilocybin and other psychedelics interact with specific "docking stations" on brain cells, influencing the flow of communication. These docking stations are called serotonin receptors, and psilocybin has a particular affinity for two types: 5HT1A and 5HT2A/C.[22] Think of serotonin as a master conductor, orchestrating various brain functions like mood, perception, and cognition. The 5HT1A receptors act like volume dials, fine-tuning the overall level of serotonin activity. When psychedelics activate these, it can lead to a calming effect and a heightened focus on internal thoughts and feelings.[23, 24]

On the other hand, 5HT2A/C receptors are more like individual instrument controls, influencing specific aspects of perception, thought patterns, and emotional responses. When psychedelics interact with these, it can explain the altered perception, intense emotions, and introspective thoughts that characterize a psychedelic experience.[25, 26] By simultaneously influencing these different receptors, psilocybin creates a unique symphony of effects, orchestrating the mental and emotional shifts experienced during a psychedelic trip.

By simultaneously quieting the DMN ("our inner voice") and amplifying the Salience Network ("our sensory filter"), psychedelics create a unique altered state of consciousness. This altered state can be associated with feelings of profound insight, mystical experiences, and a sense of unity with the universe. Things that seemed profoundly important before—like knowing you are about to die of cancer or are addicted to alcohol—no longer carry the same emotional weight, allowing us to choose new emotional states or understandings. Once the drug wears off and the trip is over, the networks return to normal—but the realizations and profound thoughts do not disappear: that is the therapeutic potential locked in a biologically safe drug.

The Ultimate Growth Mindset

So why do psychedelics have a connection to wild thinking? It feels obvious to point out by now, but the very definition of a psychedelic drug is one that allows the mind to explore new paths and have new experiences. Several factors contribute to this powerful connection between psychedelics and groundbreaking breakthroughs, but three seem to stand out.

Psychedelics loosen the rigid grip of established thought patterns, allowing for the formation of novel connections and associations. This "dislocation of the self," as some describe it, fosters a state of divergent thinking, where the mind ventures beyond the usual boundaries and explores a vast landscape of possibilities. Neuroscientist Robin Carhart-Harris has shown that psychedelics essentially cause "entropy" in the brain[27]—the drugs free our minds to a wider range of ideas, and then our minds interconnect those ideas in brand-new ways. Carhart-Harris argues that a "high-dose psychedelic experience has the power to shake the snow globe,"[28] disrupting existing patterns of thought and allowing for flexibility and creativity. This newfound fluidity often leads to the generation of innovative solutions and the birth of groundbreaking ideas as evidenced by the story of Kary Mullis and countless others who credit psychedelics with igniting their creative spark.

Beyond unlocking creative potential, psychedelics can also facilitate profound introspection and self-discovery. Under their influence, individuals often embark on a deep internal journey, prompting them to examine their own core values, beliefs, and motivations. This introspective process can lead to startling insights about the self, revealing previously unseen connections between personal experiences, the surrounding world, and even the grander scheme of existence. Additionally, some individuals report experiencing visions of the long-term self, gaining a newfound clarity about their purpose and aspirations in life. These transformative experiences can act as a catalyst for personal growth and inspire individuals to pursue moonshot goals, aiming to leave a lasting impact on the world.

Psychedelics also hold a unique power to disrupt the status quo within our minds, leading to groundbreaking ideas and paradigm shifts. They temporarily suspend the rigid mental frameworks we've built over time, the ones that often dictate how we perceive the world and limit our potential. This "ego dissolution," as it's sometimes called, allows us to challenge ingrained assumptions and societal norms that might otherwise prevent us from considering unconventional solutions.

Chris Letheby, a philosopher at University of Western Australia, even goes so far as to say that one of the powerful opportunities with psychedelics is that they allow us to "gain new knowledge of old facts"[29]—reexperiencing ideas, memories, theories, or knowledge from new perspectives, which causes us to use that knowledge in new ways. With these barriers temporarily removed, we gain the freedom to explore uncharted territories of thought, to entertain wild ideas that might have previously seemed outlandish. This process of dismantling limitations can pave the way for the birth of truly transformative concepts, as individuals are able to approach problems with a fresh perspective and unbridled creativity. In this way, psychedelics can act as a catalyst for moonshot moments, allowing individuals to dream beyond the confines of the ordinary and pursue goals that reshape the world in profound ways.

The Mushroom Doesn't Grow Far from the Tree

Our scientific understanding of psychedelics has advanced dramatically in the last twenty-five years; still, much of the research on psychedelic therapy sits at the controversial frontier of experimental study. My own interest in psychedelics began when I was prescribed psychedelic therapy as part of my own mental health journey. As part of my learning process, I picked up Michael Pollan's *How to Change Your Mind* and became fascinated, not only with the power of psychedelics in therapeutic settings, but also with the journey of psychedelic therapy research itself. I wanted to understand the complex personalities, like Kary Mullis, Timothy Leary, and others, who made up the story of psychedelics in America, as both a means by which we can produce more moonshot moments, and as a moonshot moment in psychiatric treatment itself. To gain insight into this realm, I reached out to some of the brightest minds in the field; of all those I spoke to, none captivated me as much as the father-son team of Bill and Brian Richards.

Dr. Bill Richards is a titan in the field of psychedelic research. He began his career studying religion, trying to understand how people interacted with the divine and coped with death. In the 1960s, he found himself in Europe, researching psychedelics with many of the world's leading scientists.[30] Over his career, he has developed meaningful friendships with powerhouses like Leary and traveled to funerary festivals in India, all in pursuit of a better understanding of humanity's relationship with us and our minds. His research in the late 1990s at

Johns Hopkins University with Roland Griffiths was instrumental in reigniting psychedelic therapy research in the twenty-first century. There are few humans living today who have a better understanding of the history, impact, and potentialities of psychedelics and psychedelic-assisted therapy.

Alongside his father, Dr. Brian Richards is one of today's leading psychedelic-assisted therapists, working to normalize psychedelic research and make psychedelic therapy standardized, accessible, and affordable for all Americans.[31] Drawn to psychedelic research not by his father's work, but by his own struggles coping with his mother's tragic passing at a young age,[32] Brian Richards represents the next generation of psychedelic research: focused on helping others cope with mental health struggles in an increasingly chaotic world. Together, Dr. Bill and Dr. Brian represent the narrative of psychedelic research I would read and hear about repeatedly, from the wild hippie movement through a period of prohibition to a new era of acceptance within the government and medical community.

On a rainy spring afternoon, my car dropped me off at Bill Richards' house outside Baltimore, Maryland. The pair had invited me out for a long chat, and we were able to sit down for several hours of fascinating discussion. Both men were warm and friendly, easily creating an almost meditative calm in the living room as we talked. I was struck by how passionate they were about the importance of control in the use of psychedelics, emphasizing how much of the psychedelic experience came from proper guidance from a trusted therapist who could help the individual navigate the dissolution of ego and connection to wider themes of trauma, healing, and spiritual growth.

I was quickly impressed by the focus and steadfastness of both doctors. They expressed themselves as researchers wielding powerful tools, not to be indulged in for entertainment, but to be used as keys for unlocking mental health crises of today. Both men are troubled by the fact that most pharmaceutical treatments for mental health issues are symptom-oriented rather than focused on breaking through ego and trauma to understand the experiential and social roots of mental illness. Psychedelics offer a powerful alternative with far fewer negative psychical effects.

Throughout my conversation with the Richards, the importance of "set and setting" in psychedelic therapy became a central theme. They strongly advocated for controlled settings built on trust, with trained professionals present to guide and manage the psychedelic journey. Their perspective underscored the power of psychedelics as tools that shouldn't be wielded without purpose and direction. The revelations that can surface during a psychedelic experience, especially if traumatic, can be overwhelming for those unprepared. A skilled professional can act as a guide, ensuring a safe and potentially transformative encounter. In essence, psychedelics require collaboration and hyper-cooperation to maximize the positive results of psychedelic therapy. But in this way, psychedelics can become a win-win for all.

I was also struck when hearing Brian describe the importance of standardization, regulation, and accountability when it comes to developing therapeutic treatments for mass consumption. He envisions a future when anyone in need of psychedelic therapy can access it—but a future where there are regular standards and reviews to ensure a level of care for all patients. Like all other coming wave technologies, psychedelics offer great opportunities but require us to work right now to develop the processes by which we manage and share this technology. In this way, psychedelics are yet another emergent transhumanist technology that will pave the way for more moonshot moments for humanity.

Of the many things that we discussed that afternoon, one quote lingers with me the most. I asked if there was an ideal age for psychedelic-assisted therapy; as a twenty-five-year-old myself, I knew I was on the younger end of the range of those prescribed psychedelics treatment. I expected a discussion of biology, especially around brain development, but instead, Bill responded with one of his favorite quotes drawn from the field of meditation studies:

You have to be somebody before you can be nobody.

—Jack Engler

To Bill Richards, Engler was expressing the fact that individuals must go through life experiences before having the ability to reflect on them. As Chris Letheby notes, psychedelics not only allow us to disassociate from ego and social constructs but also give us the space to reconsider knowledge we already have in new ways. Whether knowledge and experience is positive or traumatic, psychedelics give us the opportunity to think about the world and those experiences in new ways. We have to have already gained a certain amount of experience to have enough material to work with. I found this observation to be optimistic and energizing: as we age, we gain more of the wisdom that we then need to unpack (and organize in our PKM) in order to find meaning in life. Like so many other coming wave technologies, psychedelics will provide numerous opportunities for moonshot moments—if we're willing to become somebody, accept that we are nobody, and open ourselves to new frames of thought.

But how exactly does that work? I've discussed the science behind psychedelics, but for both Bill and Brian Richards, the power of psychedelics lies in the experiences and conclusions drawn during therapy, not the chemical reactions in the brain. From letting go of the self to connecting with a greater joyous cosmology to accepting death, the experience of psychedelics is a vast range we have only begun to document and study—though humans have been learning from psychedelic realization for millennia.

Substances to Open the Mind: The History of Psychedelics

As I mentioned before, much of my discussion focuses on two specific psychedelics: LSD and psilocybin mushrooms. But these two brothers are not the same age. LSD is relatively new; psilocybin mushrooms are most certainly not, boasting a rich history as powerful spiritual and healing tools. For millennia, cultures around the world have incorporated psychedelic mushrooms into their religious rituals, seeking to enhance consciousness and connect with the divine. These practices offer a powerful historical precedent, demonstrating the profound link between human augmentation through altered states of consciousness and the ability to explore unconventional, even "wild," thought patterns.

Evidence suggests the use of psilocybin mushrooms in ancient cultures for thousands of years. Cave paintings and mushroom-shaped stone artifacts found throughout Central America,[33] Europe,[34] and Africa[35] point to their prominence in both spiritual rituals and medicinal practices. The Aztecs, for example, referred to psilocybin mushrooms as "teonanácatl," translating to "flesh of the gods,"[36] and reserved their use for sacred ceremonies aimed at facilitating communication with deities and seeking guidance. The Mazatec peoples of Mexico also employed mushrooms for centuries in healing rituals,[37, 38] calling on their visionary properties to diagnose and treat ailments of both the body and spirit. Terrance McKenna, famed author and self-proclaimed "psychonaut," even went so far as to argue that mushrooms were a central part of what gave humanity the spark of consciousness: that eating mushrooms gave us myth.[39]

Perhaps humanity's greatest story wasn't fire. Maybe it's that humanity is on the greatest trip of all time.

However, the arrival of Spanish conquistadors in the sixteenth century brought about a brutal suppression of these practices.[40, 41] Seeking to establish dominance and impose Catholicism on the indigenous populations, the Spanish viewed the use of psychedelic mushrooms as a form of paganism and a threat to their control.[42] As a result, the traditional use of these substances was driven underground, a legacy that lingers even today in some communities.

R. Gordon Wasson, a renowned ethnomycologist and amateur mycologist, played a pivotal role in bridging the gap between the clandestine realm of indigenous mushroom use and the scientific curiosity of the West. In the 1950s, fueled by his fascination with the sacred rituals of the Mazatec people in Mexico, Wasson embarked on a series of clandestine expeditions, seeking guidance from a Mazatec shaman named María Sabina. Eventually he became one of the first Westerners to intentionally ingest psilocybin mushrooms during a traditional ceremony. Wasson's vivid account of the experience, published in *Life* magazine under the title "Seeking the Magic Mushroom,"[43] sent shock waves through the public consciousness, igniting widespread interest in these enigmatic fungi.

Wasson's methods and interpretations have faced scrutiny in recent years. Despite María Sabina's warnings about the sacred nature of the rituals, Wasson's published accounts ignited a firestorm. Tourists and curious celebrities flooded the remote, isolated Mazatec village of Huautla de Jimenez, disrupting their way of life. Authorities, overwhelmed by the attention and concerned about tourist behavior, accused María Sabina of drug dealing and threatened to ban the ceremonies. Her village exiled her, her house was burned down, and her son was murdered for her role in bringing outsiders to the community.[44]

Before her death, María Sabina made comments that she deeply regretted ever trusting Wasson.[45] Her openness inadvertently jeopardized her community's traditions and ruined the sacred healing nature of the Mazatec ritual. As Ben Sessa notes, "Let's not forget that when it comes to being the first to trample through jungles, whose magic is defined by their isolation from Western influence, such trail blazers are, by that very definition, also culture crushers."[46] Poorly orchestrated or not, Wasson's contribution to sparking conversations about the potential of psychedelics within the Western world remains undeniable.[47] His work paved the way for further research and exploration, laying the groundwork for the current wave of scientific investigation into the therapeutic applications of mind-altering substances.

LSD, in contrast, wasn't discovered but rather invented in a test tube. In 1938, in the otherwise ordinary halls of Sandoz Laboratories in Basel, Switzerland, a young chemist named Albert Hofmann was delving into ergot, a parasitic fungus known to grow on rye and other grains.[48] Ironically, this seemingly mundane fungus holds a dark history associated with the Salem witch trials, where the hallucinations of the girls and the mass hysteria they caused was possibly linked to accidental ingestion of ergot-contaminated food.[49] Not the best start for the psychedelic.

But while synthesizing various derivatives of lysergic acid, found in ergot, Hofmann inadvertently created a compound named lysergic acid diethylamide-25, or LSD-25. It was catalogued and put into storage, only to be brought back for tests again five years later. Thus, its effects remained unknown until April 19, 1943, when a trace amount accidentally entered his bloodstream, inducing the world's first, and rather terrifying, LSD trip. Hofmann's meticulous notes detail vivid hallucinations, a distorted sense of time, and a feeling of his consciousness leaving his body.

The full impact of Hofmann's accidental creation rippled far beyond a single, serendipitous experience. To better understand LSD's extraordinary potential, he decided to conduct a self-experiment—a controversial decision even by the standards of the time. On April 19, 1943, which later became known as "Bicycle Day,"[50] Hofmann ingested 250 micrograms of LSD (an extraordinarily high dose compared to the typical threshold today). What ensued was an overwhelming journey into the depths of his own consciousness, a transformative experience that would color Hofmann's view of his invention until his death in 2008.

Throughout the 1950s and early '60s, LSD became a focal point of psychiatric and psychological research, with countless studies examining its potential applications across a spectrum of mental health conditions. Psychiatrists like Humphry Osmond, often credited with coining the term *psychedelic*, explored the use of LSD as a tool for modeling psychosis, believing it could aid in understanding and potentially treating schizophrenia. Others, like Sidney Cohen, conducted research on the therapeutic effect of LSD for treating alcoholism, demonstrating promising results in reducing addiction and promoting lasting sobriety. Research extended beyond traditional therapy sessions, with figures such as Oscar Janiger examining its impact on the creative process, inviting artists to partake in guided LSD experiences and document their subsequent work.

Our Hippie Problem: The Politics of LSD

The research into LSD's therapeutic potential during the 1950s and early '60s wasn't contained solely within the sterile walls of laboratories and clinics. Through published articles, conference presentations, and word of mouth, news of LSD's mind-altering properties seeped beyond the scientific community, capturing the imagination of writers, artists, and intellectuals. Figures like Aldous Huxley, author of *Brave New World*, and Timothy Leary, a former Harvard psychologist, became vocal advocates for its potential to unlock creativity, enhance self-understanding, and facilitate spiritual experiences.

Timothy Leary emerged as a pivotal figure in the story of LSD and its cultural impact. Initially focused on the potential therapeutic uses, Leary conducted research with psilocybin at Harvard. However, his methods became increasingly unconventional, with reports of experiments given to students outside of controlled settings. This, coupled with his growing public persona as a psychedelic cheerleader, clashed heavily with societal norms. Leary's now-famous slogan, "turn on, tune in, drop out," was seen by many as an endorsement of drug use and a rejection of mainstream values. His dismissal from Harvard and subsequent arrest for marijuana possession only fueled the flames of the debate. Leary became a counterculture icon, seen by some as a visionary exploring consciousness and by others as a dangerous figure undermining public safety and social order.

As LSD's popularity surged throughout the 1960s and its use became intertwined with the burgeoning counterculture movement, a collision course with the political establishment became inevitable. President Richard Nixon, ascending to power on a platform of "law and order," viewed the rise of drug use, particularly psychedelics, as a direct threat to "conservative American values" and social stability. He saw the psychedelic movement as inextricably linked to a growing anti-war sentiment regarding the Vietnam War, which was fueling public unrest and fueling divisions within American society. As scientist Ben Sessa explains:

> The role of the Vietnam War and the protests against it in the development of the psychedelic scene is undeniable. What better antithesis could there be of the LSD experience than a brutal televised war that dragged teenagers kicking and screaming away from their books and music and dropped them into the dark jungle thousands of miles away from home? If it is an intrinsic and spontaneous characteristic of an LSD trip to feel the love, then people now had many more reasons to strive for it.[51]

In this context, Nixon declared the War on Drugs, a concerted governmental effort to crack down on the production, distribution, and consumption of illicit substances. This campaign targeted not only street drugs but also psychedelics, with Nixon famously labeling Timothy Leary "the most dangerous man in America." The implementation of the Controlled Substances Act in 1970[52] classified LSD, psilocybin, and other psychedelics as Schedule I substances—deemed to have no medical value and a high potential for abuse. This was a particularly hypocritical position for the U.S. government to take, considering that the CIA had been giving psychedelics to unknowing test subjects for twenty years as part of their own Project MKUltra.[53, 54]

Scientist Robin Carhart-Harris once joked to journalist Michael Pollan, as they talked about psychedelics' ability to overturn mental and social hierarchies: "Was it that hippies gravitated to psychedelics, or do psychedelics create hippies? Nixon thought it was the latter. He may have been right!"[55] And, indeed, Nixon may have been right: according to research by Matthew Nour and Carhart-Harris, "ego dissolution experienced during a participant's 'most intense' psychedelic experience positively predicted liberal political views, openness and nature relatedness, and negatively predicted authoritarian political views."[56] If ever there was a combination of authoritarian policies to rebel against, Nixon's Vietnam War draft and his War on Drugs would fit the bill.

Besides being politically motivated and inherently racist, this War on Drugs, under Nixon's administration, changed the way much of the public perceived psychedelics. This prohibition not only hampered the exploration of their therapeutic potential but also cast a shadow of criminality and social stigma over these substances. The echoes of these decisions still reverberate today, shaping public perceptions of psychedelics and influencing the complex regulatory landscape that governs their research and potential reintegration into medical practice.

Scheduling psychedelics as Schedule I substances in 1970 proved to be a formidable roadblock to scientific progress. This classification, based on the belief that these drugs have no accepted medical use and high potential for abuse, made research immensely difficult and discouraged potential investigators. The bureaucratic hurdles, immense costs, and potential damage to careers associated with acquiring Schedule I drugs for research deterred many scientists from pursuing this avenue. This restrictive policy effectively halted most research on psychedelics for decades, hindering the exploration of their potential therapeutic benefits and delaying the understanding of their mechanisms of action. The consequences of this scheduling decision continue to be felt today as researchers navigate complex regulatory frameworks and strive to overcome the stigma and misinformation surrounding these once-promising therapeutic tools.

The history of psychedelic research and its eventual criminalization serves as a stark reminder of the potential pitfalls of overreacting to new technologies and substances that challenge conventional norms. If the burgeoning fields of AI and synthetic biology are perceived as uncontrollable or disruptive by governments of the future, there's a genuine risk of them facing similarly harsh regulations and censorship. Knee-jerk reactions driven by fear or a lack of understanding could stifle critical innovations and breakthroughs that could potentially benefit society. This underscores the importance of engaging in responsible research, proactive public education, and open dialogue to shape the development of these powerful technologies in a way that maximizes their potential benefits while addressing potential risks in a measured, informed fashion.

Psychedelics in Silicon Valley: From Counterculture to Moonshot Medicine

While the direct causal link is difficult to establish definitively, the unique counterculture scene that flourished in San Francisco, Berkeley, and Stanford during the 1960s undeniably played a role in laying the groundwork for the rise of Silicon Valley. This era was characterized by a spirit of experimentation, a rejection of traditional authority, and a fascination with exploring the potential of the human mind.

Psychedelics, readily available and viewed by some as tools for personal growth and creative exploration, became intertwined with this cultural zeitgeist.[57] Figures like Ken Kesey, who wrote *One Flew Over the Cuckoo's Nest*, in part, while on psychedelics, explored LSD's potential for personal growth and societal change. Visionaries like Douglas Engelbart, considered the father of the computer mouse, participated in research at the Stanford Research Institute using LSD to enhance creativity and problem-solving in the early 1960s.[58]

However, as the counterculture faded and psychedelics became demonized in the 1970s, their presence in Silicon Valley receded into the background. It wasn't until the late 1990s that interest rekindled, fueled by figures like Paul Stamets, a mycologist and advocate for the potential of psilocybin mushrooms. Stamets, despite lacking a traditional PhD, is a fascinating figure in the world of mycology. His lack of formal credentials hasn't hindered his impact; he is widely considered one of the most respected and influential mycologists of our time.[59] He has authored numerous books, founded Fungi Perfecti, a company dedicated to the cultivation and study of mushrooms, and spearheaded countless research projects. His combination of intellectual rigor and "do it yourself" attitude has made him a folk hero, not only in the field of psychedelic research, but also among Silicon Valley founders who see him as a kindred spirit.

Stamets's work, alongside the emergence of organizations like the Multidisciplinary Association for Psychedelic Studies (MAPS) in San Jose, founded by Rick Doblin, helped pave the way for a more scientific and responsible exploration of psychedelics. Thanks to respected psychologist James Fadiman's exploration of microdosing[60] (taking small doses of psilocybin to inspire creativity), a new generation of tech titans started exploring altered states of consciousness in search of the next big breakthrough. By 2015, CNN was asking if taking LSD would actually make you a millionaire, as it was so common for tech titans to drop acid.[61]

Today, Silicon Valley is once again abuzz with psychedelic interest. Many tech entrepreneurs and investors have acknowledged the influence of psychedelics on their own lives and careers. While not advocating for recreational use, it's noteworthy that Steve Jobs, the cofounder of Apple, openly discussed his use of psychedelics in his younger years. He called taking LSD "one of the two or three most important things I have done in my life,"[62] and believed it reinforced "his sense of what was important—creating great things instead of making money."[63] Jobs credited psychedelics with helping him "break out of traditional ways of thinking" and fostering a creative mindset that fueled his entrepreneurial spirit. He even reportedly told Bill Gates that Microsoft would be a better company if Gates "dropped acid."[64]

Venture capitalist and PayPal cofounder, Peter Thiel, has become a prominent figure in the resurgent interest in psychedelics. Thiel has invested heavily in the field, backing Atai Life Sciences, [65, 66, 67] a German company developing psilocybin-based therapies, with the belief that these substances can "unlock the full potential of the human brain." This investment underscores his conviction in the therapeutic value of psychedelics, potentially shaping the future of their development and accessibility.

The conversation around psychedelics in Silicon Valley extends beyond theoretical discussions and historical influences. Prominent figures in the tech industry are actively engaging with these substances. Tesla,[68] SpaceX, and OpenAI founder Elon Musk has openly discussed his experimentation with microdosing ketamine, [69, 70] while Google cofounder Sergey Brin reportedly dabbles in small amounts of psilocybin mushrooms.[71]

Psychedelics are becoming an intriguing influence on the mindset of those shaping the future of AI, with figures like Sam Altman, CEO of OpenAI, embracing them as tools for personal transformation and creative exploration.[72] Altman, who describes himself as a formerly anxious and unhappy person, credits a guided psychedelic retreat in Mexico with profoundly shifting his perspective, leaving him calmer and more capable of managing the immense challenges of developing artificial general intelligence. His experiences at Burning Man gave him glimpses of communal joy and creativity that resonated deeply, which he described as "one possible part of what the post-AGI world can look like."[73] While Altman has explored psychedelics in guided settings, he also values the potential of sober reflection and sees these experiences as medicine— offering insights not only for personal growth but for imagining the future of human-AI collaboration.

Even beyond individual experimentation, venture capital firms like Founders Fund, known for investments in major players like SpaceX and Facebook, have hosted events where psychedelics are reportedly present.[74] These anecdotes point to a growing openness and exploration of psychedelics within the heart of Silicon Valley, suggesting a potential shift in attitudes and a willingness to explore the potential benefits and risks of these substances in both personal and professional spheres.

Finding Hope: A Moonshot for Mental Health

Following the decades-long lull from the strict regulations of the 1970s, a new wave of scientific exploration is gaining momentum, driven by a confluence of factors.

First, mounting evidence from anecdotal reports and preliminary research hinted at the potential therapeutic benefits of psychedelics for various mental health conditions, including depression, anxiety, and addiction. Second, advancements in brain imaging technology allowed researchers to noninvasively examine the effects of psychedelics on the brain during the drug experience, shedding light on their potential mechanisms of action.[75, 76]

Third, artificial intelligence is revolutionizing psychedelic medicine, transforming drug discovery, clinical trials, treatment personalization, and post-treatment integration.[77] Researchers have used AI, such as the AlphaFold program, to identify hundreds of thousands of new psychedelic compounds, potentially leading to medications that retain therapeutic benefits while minimizing side effects.[78] In clinical trials, AI analyzes patient data to optimize dosing strategies, predict treatment efficacy, and monitor real-time outcomes with wearable technology. It also offers deeper insights into the psychedelic experience by uncovering patterns in therapy session data, which can inform more effective treatments.[79] Tools like virtual reality and personalized sensory inputs further enhance therapy by creating tailored environments to amplify healing. As AI continues to drive these advancements, ethical considerations remain crucial, ensuring patient data security and responsible technology use in this sensitive and transformative field.

To learn more about the mental health moonshots that are emerging from psychedelic therapy research, I sat down with Dr. Rick Doblin, founder of the Multidisciplinary Association for Psychedelic Studies (MAPS). Over the course of three hours of interviewing, Rick shared insights on the process of legalizing psychedelics for research, and the massive breakthroughs on the horizon. Since founding MAPS in 1986, Doblin has worked tirelessly to advance the therapeutic use of psychedelics, most notably MDMA for PTSD treatment, with clinical trials yielding groundbreaking results.

Easy to chat with and energized by recent shifts in governmental perceptions of psychedelic research, Doblin painted a picture of a varied research landscape, with advances made with some drugs while others still struggled to find governmental support. Ibogaine, which featured in many studies with combat veterans,[80] was easily finding support on both sides of the aisle, with psychedelic-adjacent substances like ketamine and MDMA also gaining more traction than their hallucinogenic counterparts. Still, diligent efforts and partnerships between academia, non-profits, emerging startups, and the therapeutic community are continuously necessary to convince the federal government of the importance of legalizing guided psychedelic therapy and research.

When activists like Rick Doblin encourage increased research into the psychiatric uses of psychedelics, it's important to note that this research is conducted in highly controlled settings with trained professionals and strict protocols to ensure participant safety and ethical conduct—just as the Richards team recommended. In the realm of psychedelic research, the concept of "set and setting," first introduced by Leary in his work,[81] reigns supreme.[82] This goes beyond simply having a comfortable couch and good music—although that definitely helps make the experience better! Set refers to the mental and emotional state a person brings to the experience. Factors like anxiety, past trauma, or even recent life events can significantly influence the trip. Likewise, setting encompasses the physical and social environment. A quiet, supportive space with trusted companions fosters a safe and positive experience, while a chaotic or unfamiliar setting can amplify anxieties and lead to a negative trip. These factors aren't just background noise; they're integral ingredients that can shape the entire journey. Researchers meticulously control both set and setting to ensure participant safety and maximize the potential for therapeutic breakthroughs.

Psychedelics assist in transpersonal psychology, which focuses on how understanding the spiritual and transcendental aspects of the human experience shape how we interact with the world.[83] Studies have demonstrated the potential of psilocybin therapy in treating treatment-resistant depression,[84] anxiety associated with terminal illness, and even end-of-life distress.[85] Psychedelics have also been shown to help with OCD patients,[86] who are able to reprioritize their fears and move past them.

Psychedelics are pivotal in treating addiction; for example, there has been significant progress in the field of alcohol addiction with the use of psilocybin therapy.[87, 88] Psychedelics have helped smokers reduce use or kick the habit entirely.[89, 90] Considering the United States is currently in the throes of an opioid epidemic, it's very exciting to report that psychedelics, as well as ketamine, have been shown to be tremendously effective in treating heroin addiction.[91, 92, 93]

There has also been incredible work treating U.S. military veterans with PTSD with psychedelics,[94] a field of research that I believe our federal government has a moral imperative to fund and support. Several outstanding organizations are dedicated to helping veterans, including Veterans Exploring Treatment Solutions (VETS), founded by retired Navy SEAL Marcus Capone.[95] Capone, who served thirteen years in the military and faced significant mental health challenges post-service, discovered the transformative potential of psychedelic-assisted therapies. In response, he and his wife Amber established VETS in 2019 to support veterans struggling with PTSD and other mental health conditions. VETS provides resources, education, and funding for veterans seeking psychedelic therapies, advocating for their safe and effective use to foster healing and improve lives.

The path toward the legalization of psychedelics is being paved by groundbreaking work, particularly with veterans. Ibogaine, an African-origin psychedelic, is emerging as one of the frontrunners in this effort. Thanks to organizations like VETS that facilitate treatments in Mexico,[96] a structured protocol combining ibogaine and 5-MeO-DMT is showing extraordinary promise in treating PTSD. This dual approach involves intensive therapy leading up to the experience, ceasing all prescription medications, taking ibogaine, waiting forty-eight hours, and then taking 5-MeO-DMT.[97] The result is a powerful therapeutic reset, especially for those grappling with severe PTSD or addiction. The success of these protocols highlights how tailored and intensive interventions can offer hope for some of the most challenging cases.

Recently, we have a great example of the kind of academia-government collaborations we need more of: in late 2023, Congress passed legislation providing funding for clinical trials of psychedelic-assisted therapy for active-duty service members to treat anxiety, PTSD, and depression.[98] At the beginning of 2024, the Department of Veterans Affairs announced that it will also begin funding psychedelic-assisted therapy to treat veterans with PTSD and depression.[99] Maybe therapists can help our remaining Vietnam veterans finally find their peace, despite all of Nixon's efforts to the contrary.

If ever there was a time that we need some moonshot solutions for mental health issues, it's now: we're currently facing a mental health crisis unlike any other. Rates of anxiety and depression are soaring, with the World Health Organization estimating that one in eight people globally live with a mental health disorder.[100] In the United States alone, more than forty million adults experience anxiety disorders each year, and suicide remains a leading cause of death, particularly among young people.[101]

Traditional treatments haven't always kept pace with the rising tide of mental health struggles, highlighting the urgent need for new and innovative approaches. While valuable, they sometimes feel like rearranging deck chairs on a sinking ship.

We need a moonshot—a groundbreaking approach that could revolutionize how we treat mental illness. Solving mental health issues is a space where we should first try psychedelics. We can already see a future in which psilocybin therapy helps veterans overcome debilitating PTSD, or LSD-assisted sessions provide a reset button for those trapped in the cycle of depression. Psychedelics aren't a magic bullet, but they offer a glimmer of hope, a potential moonshot that could empower doctors to finally help the millions struggling worldwide find a path back to well-being.

While psychedelics hold promise for the current mental health crisis, the future landscape might present entirely new challenges. As AI automation takes over routine jobs, vast segments of the population could face an existential crisis. Stripped of the structure and purpose traditionally provided by work, individuals might grapple with feelings of purposelessness and a void in their daily lives.

For some, the prospect of boundless free time might seem like a utopia. But the reality could be far bleaker. The abundance of free time, once a dream, could morph into a burden, leading to increased social isolation, boredom, and a potential rise in mental health issues stemming from a lack of identity and fulfillment. Unstructured time, devoid of purpose and filled with a constant hum of existential questions, would easily become a breeding ground for increased anxiety and depression. The key lies in transforming this free time from a burden into a gift, a chance to unlock the full potential of the human mind.

Used responsibly and within controlled settings, these substances have shown promise in facilitating a state of heightened curiosity, introspection, and a loosening of rigid thought patterns. How many polymaths, unburdened by working to survive, could use psychedelics to ask new philosophical questions, build unimaginable inventions, create breathtaking art, or solve the greatest problems of their day?

By temporarily quieting the inner chatter and amplifying the connections between disparate ideas, psychedelics (particularly when paired with mindfulness and meditation practices) can open doors to new ways of thinking and experiencing the world. This isn't about escaping reality; it's about confronting it with a renewed sense of wonder and possibility.

I believe this newfound capacity for exploration could lead to a future brimming not just with solutions to our current problems but with entirely new ideas and innovations across all fields. In essence, psychedelics might not just solve our mental health crisis; they could be the key to unlocking a future overflowing with moonshot moments that propel humanity forward.

Embracing the Joyous Cosmology

Psychedelics can orchestrate a symphony of altered perception, transforming the world into a kaleidoscope of vibrant colors, sounds, and sensations. Psychedelics can induce a fascinating phenomenon known as synesthesia, where the boundaries between our senses blur. Have you ever had the experience of hearing a song and not just perceiving the sound but also tasting a burst of citrus or seeing a kaleidoscope of colors dance before your eyes? It sounds like a recipe for a Willy Wonka treat, but synesthesia happens because psychedelics act as a bridge, temporarily connecting brain regions that don't typically communicate directly. This "cross-talk" can manifest in a multitude of ways, from associating specific sounds with colors to experiencing emotions as textures. The world becomes a symphony of interconnected sensations, offering a unique and often overwhelming sensory experience that can be both challenging and awe-inspiring. This heightened sensory awareness can also extend to the way we perceive time and space.

Time may seem to slow down or even bend while the boundaries between objects might appear to blur, creating a sense of interconnectedness and a deeper appreciation for the intricate details of the world around us. Alan Watts, a famous psychedelic and Zen Buddhist philosopher, gives us a visceral explanation:

> To begin with, this world has a different kind of time. It is the time of biological rhythm, not of the clock and all that goes with the clock. There is no hurry. Our sense of time is notoriously subjective and thus dependent upon the quality of our attention, whether of interest or boredom, and upon the alignment of our behavior in terms of routines, goals, and deadlines.[102]

This distortion can be understood through that Default Mode Network (DMN) we discussed before. This network, typically active when we're lost in introspection and self-referential thought, serves as our internal timekeeper, weaving together memories, plans, and the present moment into a coherent narrative of our experience. So when psychedelics quiet the DMN, disrupting its usual symphony of self-referential chatter, we can't actually tell if time has passed.[103] Without this internal anchor, our sense of time becomes fragmented. External stimuli take on a heightened presence, potentially stretching or compressing subjective time.

Ego dissolution, a hallmark experience of some psychedelic journeys, transcends the simple feeling of "losing oneself." It's a temporary state where the rigid boundaries we usually construct around our sense of "self" soften and dissolve. This can lead to profound feelings of interconnectedness, a sense of being one with everything around us, and a blurring of the lines between self and environment. Imagine viewing the world, not as a collection of separate objects, but as a unified whole, where you are not just an observer but an integral part of the tapestry of existence. Alan Watts describes this as communing with the "Joyous Cosmology," a phrase that is now used by many in the realm of psychedelic writing and research to refer to the larger "force" that those under the influence connect to. This experience can be deeply transformative, fostering feelings of profound empathy, a sense of universal belonging, and a newfound appreciation for the interconnectedness of all things. However, it can also be challenging, as it can confront us with the vastness and impermanence of our individual selves, potentially leading to feelings of anxiety or a temporary sense of disorientation.

One intriguing effect of psychedelics lies in their potential to unlock forgotten memories tucked away in the subconscious. Our brains are constantly filtering and prioritizing information, often relegating past experiences to the depths of our minds. Psychedelics, however, may act as a key, quieting the usual analytical chatter that governs our waking consciousness. This allows access to deeper emotional landscapes and personal narratives, potentially bringing to light long-dormant memories. These memories can range from joyful childhood experiences to repressed emotional wounds. While surfacing these memories can be emotionally challenging, it can also be a catalyst for healing and personal growth. By integrating these forgotten aspects of ourselves, patients can gain a deeper understanding of who they are and move forward with a more complete sense of self. As these memories can sometimes be traumatic, it's important to remember that this process should be undertaken with caution and with the guidance of a qualified professional to ensure a safe and supportive environment.

Beyond unlocking buried emotions and fostering self-compassion, psychedelic therapy can also cultivate a profound sense of interconnectedness with the world around us. This isn't just a warm fuzzy feeling, but a radical shift in perspective that could be crucial for fostering hyper-cooperation in the future. Under the influence of psychedelics, many users report feelings of dissolving boundaries between self and other, fostering a deep sense of empathy and understanding. This allows them to see the world through a "win-win" lens, a world where cooperation and collaboration become intuitive acts based on a shared sense of well-being for all. In fact, Arce and Winkelman[104] argue that prehistoric humans' hyper-cooperative nature came from eating psychedelic mushrooms: perhaps psychedelics are the ultimate key to our hyper-cooperative future. Psychedelics thus could act as powerful tools for fostering the kind of collective spirit we'll need to tackle the complex challenges of a globally interconnected future.

However, it's crucial to note that these experiences are subjective: some people feel a connection to a religious divine, some to the joyous cosmology or to some broader force of the universe, and some to just a broader place in their own personal existence. In other words, there's not one single spirituality or divinity that holds for psychedelics. Michael Pollan describes this well, stating:

> I have no problem using the word "spiritual" to describe elements of what I saw and felt, as long as it is not taken in a supernatural sense. For me, "spiritual" is a good name for some of the powerful mental phenomena that arise when the voice of the ego is muted or silenced. If nothing else, these journeys have shown me how that psychic construct—at once so familiar and on reflection so strange—stands between us and some striking new dimensions of experience, whether of the world outside us or of the mind within. The journeys have shown me what the Buddhists try to tell us but I have never really understood: that there is much more to consciousness than the ego, as we would see if it would just shut up. And that its dissolution (or transcendence) is nothing to fear; in fact, it is a prerequisite for making any spiritual progress.[105]

Rather, there are universals across faiths and spiritualities, moments of wisdom that connect all of humanity. Psychedelics advocate Aldous Huxley's seminal work, *The Perennial Philosophy*,[106] shows its readers the core similarities shared by diverse world religions. A prominent theme Huxley presents is the universal human quest for transcendence—a state of consciousness beyond the ordinary, where individuals connect with something larger than themselves. Transcendence is an open celebration of empathy and a profound shift in perspective: an awakening to a deeper reality. Interestingly, the psychedelic experience mirrors this religious pursuit. By altering consciousness, psychedelics induce states resembling mystical experiences, offering a glimpse into a transcendent realm even for those like Pollan who do not subscribe to a particular mantra. This convergence of timeless spiritual practices and modern scientific exploration has profound implications.

Finally, psychedelics can act as a mental reset button, leading to a changed prioritization of ideas within the mind. This manifests in a remarkable ability to loosen the grip of ingrained assumptions and cultural norms, allowing individuals to experience and accept new realities with an open mind. Imagine the rigid scaffolding of our thought patterns temporarily dissolving, allowing previously marginalized ideas and perspectives to come to the forefront. This can lead to a reevaluation of long-held beliefs, a dismantling of self-imposed limitations, and a newfound openness to possibilities previously deemed unthinkable.

It's important to understand that this experience isn't akin to brainwashing or abandoning all critical thinking. Instead, it's a temporary suspension of judgment, allowing us to view the world through a fresh lens and consider alternative perspectives without the usual filters and biases. This can lead to a deeper understanding of ourselves, our place in the world, and the interconnectedness of all things. However, it's crucial to reintegrate these insights into our existing framework upon returning to a normal state of mind, critically evaluating the newfound perspectives and integrating them into a more nuanced and well-rounded worldview.

Mushrooms Growing on the Path to Enlightenment

Psychedelics belong to the group of substances known as entheogen, meaning "showing the divine within." Entheogens cause feelings of interconnectedness, a dissolution of ego, and a deep sense of awe and wonder at the mysteries of existence. Psychedelics have the power to initiate deep mystical or spiritual experiences and provide windows into a world beyond the mundane. These experiences can take many different forms, ranging from an indescribable realization of the interconnection of all things to emotions of a strong connection to a higher power or a sense of unity with the universe. Imagine being overcome with a deep sensation of amazement and wonder, sensing the presence of love, or suddenly realizing how interrelated everything is. These experiences, while difficult to articulate in conventional language, can be deeply transformative, fostering feelings of peace, acceptance, and a renewed sense of meaning and purpose in life.

For thousands of years, indigenous cultures across the Americas have incorporated entheogens, or psychoactive plants, into their religious practices. As I've already described, psilocybin mushrooms played an integral part in numerous religious rituals within Aztec and Mayan cultures in central Mexico, as well as in smaller rural communities across the Yucatan. Similarly, peyote cacti, containing the psychoactive compound mescaline, have been central to rituals among some North American tribes for visions, spiritual connection, and communication with ancestors.

Deep within the lush Amazon rainforest, the tradition of ayahuasca has endured for centuries among numerous indigenous tribes.[107] This potent brew, prepared from the *Banisteriopsis caapi* vine and other plant mixes, holds immense significance as both a sacred medicine and a gateway to the spiritual realm. For tribes like the Shipibo-Conibo of Peru[108] or the Yawanawá of Brazil,[109, 110] ayahuasca ceremonies are deeply ingrained in their cultures. Shamans, or *curanderos*, skillfully lead these rituals, utilizing the visionary effects of ayahuasca to commune with spirits, seek guidance, and diagnose and treat physical and spiritual ailments.

Even beyond the Americas, historical evidence suggests the use of entheogens in ancient religious practices. Soma, a mysterious "pressed juice" revered in the *Rigveda*, the oldest religious text of Hinduism, is believed by some scholars to have been derived from psychoactive mushrooms or plants. Although the exact nature of Soma remains debated, its presence in sacred texts highlights the potential role entheogens played in shaping spiritual experiences across continents and throughout human history.

There is also strong evidence that the consumption of psychedelic entheogens shaped ancient Greek and Christian traditions. For almost two millennia, academics have been perplexed by the quixotic "Mysteries rituals" held at Eleusis in Greece, which involved odd stories of initiates going on journeys beyond the physical realm. In a coauthored work, three top experts—psilocybin Westernizer Gordon Wasson, LSD inventor Albert Hofmann, and historian Carl Ruck— argue that the sacred drink administered to Mysteries participants during the rite contained a psychoactive entheogen.[111, 112] Brian C. Muraresku builds upon their thesis, confirming the importance of psychedelics in ancient Greek rituals before arguing that this tradition carried on into early Christian mysticism as the Eucharist—the very drinking of the wine and blood of Christ.[113] John Allegro similarly draws connections between mushroom use in ancient Fertile Crescent civilizations and the practice of eating the Eucharist in Christianity, pointing out how often mushrooms are depicted in medieval Christian art.[114]

Both modern religious philosophy and secular spiritualities have the ability to embrace psychedelic exploration. William Richards contends that psychedelic usage can promote a wider appreciation of various spiritual viewpoints and enhance religious comprehension. Richards claims that psychedelics enable people to deeply explore and actualize their own religious views while also developing a deeper appreciation for other people's perceptions of the divine.[115] He contends that religion is essential to society because it strengthens social ties and offers consolation in the face of death. He argues:

> Many people today, perhaps increasing in number, express their dissatisfaction with current institutional forms of religion by describing themselves as "spiritual, but not religious." Some tend to abandon the social expression of spirituality in cohesive religious communities and perhaps unwittingly allow the word "religion" to come to pertain only to the institutional, dogma-centered manifestations that they may view as lacking in inspiration, vitality, and

relevance in the modern scientific world. If indeed the word "religion," originating in the Latin *religare,* is to continue to signify that which most profoundly binds us together and reflects a shared perspective on what gives life its deepest purpose and meaning, I personally do not support abandoning it in favor of "spirituality."[116]

As such, Richards thinks that upholding a theological framework is critical to the welfare of the individual as well as the larger society, but that psychedelics can allow a mentally healthy acceptance of the greater truths of the religious experience.

Others see psychedelics as an opportunity to finally eschew unnecessary religious supernaturalism. Chris Letheby, a naturalist philosopher, approaches psychedelics from a nonreligious and nonspiritual standpoint, yet still considers the experience to be deeply spiritual. For Letheby, the value of psychedelics lies in their ability to provoke challenging self-reflection and encourage new ways of thinking about knowledge and existence.[117] While he does not believe in religious or supernatural phenomena, Letheby argues that psychedelics compel individuals to confront profound questions about themselves and their place in the world, and thus, "psychedelic research shows that there are genuine forms of spirituality that do not rely on non-naturalistic physics."[118] He argues that psychedelics are actually a form of "moral bioenhancement"—we gain a better understanding of our morals and values through their use, essentially upgrading our moral selves, even outside of the context of religious practice.[119]

Michael Pollan, who initially rejects both organized religion and Watts's Joyous Cosmology, acknowledges the profound sense of connection to a greater unity and reality that psychedelics can facilitate. He emphasizes that psychedelic spirituality can be particularly valuable for individuals grappling with the concept of death, and his anecdotal descriptions of his own trips suggest that he found profundity without prophets and that sacredness is derived from within through contemplation.[120] Pollan's experiences suggest that psychedelics offer a means to confront and reconcile with mortality, providing a sense of peace and understanding that operates as an alternative to religious models of the afterlife.

Thus, psychedelics already have the ability to change our conceptions of the religious and the spiritual. As I discussed in Chapter Four, we stand at a moment of profound technological change and spiritual tumult. Traditional religions may struggle to adapt to the rapid shifts in our understanding of life, death, and our place in the universe. This is where I believe that entheogens, like psychedelics, offer a fascinating and potentially transformative tool. Let us embrace responsible research and utilize these tools thoughtfully and ethically.

By integrating the insights gleaned from entheogenic experiences with the wisdom of our ancestors and the advancements of science, we can cocreate meaning systems that resonate with the complexities of our ever-evolving world. Put another way: if AI and transhumanism makes it hard to find God, maybe psychedelics can help our minds bridge the gap. This is not about abandoning past values or meanings, but about fostering a future where science and spirituality can coexist and inform each other, ultimately leading to a deeper understanding of ourselves, one another, and the universe we inhabit.

My Own Trip Down the Rabbit Hole

Much like Michael Pollan (whose work first inspired me to take psychedelics seriously) and other curious minds who have ventured into the world of psychedelics, my exploration extends beyond the theoretical. Having personally experienced psychedelic substances within controlled, therapeutic settings allows me to offer not just factual information but also a glimpse into the subjective nature of the psychedelic journey.

In January 2022, I embarked on this path for a specific reason. As a multi-company founder, I felt burnt out, uninspired, and far from the passionate entrepreneur I once was. Further compounding the issue, I had been diagnosed with PTSD and depression, stemming from childhood trauma as well as an unhealthy work-life balance. To stay afloat and keep the momentum going, I had been on Adderall after college to quell my ADHD symptoms, but after a year, the medication felt like it was hurting my overall mental health more than helping. I had thrown myself into my work but felt stretched thin, unable to enjoy the parts of work and life I once loved, and ready for some kind of big change.

My therapist at that time was proving ineffective, and my psychiatrist's recommendation of adding more pills to the mix was unappealing. Intrigued by the rising culture of psychedelics and psychedelic therapy clinics in Los Angeles, I was ready to try something new and outside of "Big Pharma." I decided to pay a visit to a psychedelic therapy clinic called Stella.

My experience at Stella served as an ideal introduction to psychedelics. I learned from Stella how to treat psychedelics with respect in such a way that you can have meaningful spiritual and psychological shifts that change your outlook on life in profound ways. When done in the right setting with the right mindset, ideally with a therapist or guide who is acutely aware of the unique medical realities of psychedelics, psychedelic experiences can make you believe in the magic of the universe in ways you likely have not experienced since you were a child. In a world where the miraculous has been replaced with technology, psychedelics still offer a chance to tap into the magical, like Alice in Wonderland, chasing white rabbits.

The experience of being on psychedelics is rather ineffable, but I'll try to give an example. I remember one particular session in which I was confronted with a younger version of myself. He was so proud of the person I had become today and all I had accomplished. But that younger version of myself was accompanied by all the friends and people who I had lost along the way of becoming the person I am today. In a therapeutic setting, you do your best to vocalize what you're experiencing, seeing, and feeling so that the therapist can jot it down and help you make sense of it in the following days, through the process of integration. I later came to find that my vision of lost friends and family was some subconscious survivor's guilt, holding me back from being grateful for the life I'm lucky to have today.

Each psychedelic journey teaches me something different. Sometimes, I'm able to see and appreciate the duality of all the moments I would otherwise view as depressing moments, failures, or losses. For an extended period of time, I am able to sit and be really present with memories, thoughts, and feelings of mine, and to see them with a new lens, a new perspective, and even new sensations inside my body. When talking to friends, I often liken the time I'm on psychedelics to sitting by a fire of colors.

I become recharged with curiosity, wonder, deep philosophical thought, and belief in some sort of unity within the human experience. The level of empathy, self-compassion, and gratitude I gain for myself from this work propels me with an afterglow of energy to go and build a greater world for myself, make changes in my life, and strive for more balance in my life.

Psychedelics allowed me to shed the weight of my ego and to view myself with an objective and compassionate lens. For the first time, I could see all of the challenges and pain points that lay deep in my subconscious, limitations I created for myself, memories that I locked away, and thoughts I was too scared to reconcile; I could see and experience them now without judgment, as if observing them from a distance. This newfound perspective unlocked a deep well of self-compassion, allowing me to disconnect from the emotional pain that had held me captive for so long. It was a profound and transformative experience, a journey of self-discovery bathed in the ethereal glow of connection and understanding.

Psychedelic therapy has been a transformative journey for me, allowing me to create awareness and a new relationship with hidden patterns, memories, and emotions that shape who I am. One of the most profound early experiences involved confronting my long-held angers and anxieties. During a session, with the guidance of a therapist and the carefully crafted setting, these deeply rooted patterns surfaced.

But unlike traditional therapy, psychedelics allowed me to approach these memories with a newfound sense of love and empathy. I could finally acknowledge and grieve the emotions my younger self had neglected, emotions that often manifested as anger in my adult life. Revisiting these experiences wasn't just about understanding the past, it was about gaining control over the emotional triggers in the present. Now, when similar memories or feelings arise, I meet them with calmness and understanding. Instead of repressing them as negative aspects of myself, I recognize them as part of the story that made me who I am today. I can embrace both the positive and negative, finding peace with the experiences that shaped me.

A significant breakthrough involved the realization that my ego was hindering my success. My desire to be a "lone wolf" CEO was actually detrimental to both my companies and my well-being. Recognizing this allowed me to make crucial changes, empowering new leaders and transitioning my ventures into a collaborative incubator.

Psychedelic therapy hasn't been a quick fix but rather a powerful tool for self-discovery and growth over the last two years. By offering a unique perspective on my past and present, it has empowered me to make positive changes in my life and become the best version of myself. Psychedelic therapy has been a catalyst for personal growth unlike anything I've ever experienced. It's opened doors to self-awareness and emotional healing that traditional therapy simply couldn't reach. While I wouldn't advocate for recreational use—the "set and setting" are crucial for meaningful journeys—I believe there's immense potential here.

Controlled dosing and guided trips focused on an internal transformation could not only resolve psychological struggles but also unlock creative problem-solving and innovation. As future technologies rapidly evolve, the existential crisis of purpose looms large. Psychedelic-assisted therapy might be the key not just to understanding ourselves but also to navigating the complex questions of meaning and purpose that lie ahead. Investing in responsible research is an investment in the future—the future of mental health, creativity, and even the very soul of humanity.

Last chapter, I explained how we make ourselves more productive; with psychedelics, we would also become more open-minded. And as we embark on this journey of unlocking human potential through psychedelics, it becomes clear that this is merely the first chapter in a larger story. Transhumanism, in its vast embrace of technological and scientific advancements aimed at human enhancement, offers a multitude of avenues to explore. Having unlocked the potential of psychedelics for individual and collective transformation, let's now explore the possibilities transhumanism offers as a tool for spawning a species-wide movement toward moonshot moments.

Chapter Eight

A CALL FOR DEMOCRATIC TRANSHUMANISM

Biohacker and ex-Braintree Founder Bryan Johnson is a man who thinks he can live forever. Johnson doesn't traffic in fad diets or mindfulness mantras: his lifestyle is a full-on rebellion against the tyranny of time.

Through his Project Blueprint, a self-funded experiment, Johnson aims to prove that science can slow down, or even rewind, the aging process. He believes consistent interventions can "de-age" his body at the cellular level, restoring it to a younger state. His days are a kaleidoscope of cutting-edge interventions. Think cryotherapy chambers that leave you invigorated instead of shivering, personalized light therapy panels that bathe you in the glow of youth, and even the occasional blood transfusion, just for a touch of intrigue.

Let me give you a sample of how he describes his own days—a schedule I couldn't even make up if I tried!

> Johnson often says that he has built an algorithm that takes better care of him than he can take care of himself. Although the day-to-day details vary, he generally wakes up around 4:30 AM, and

the measurements begin: body fat, muscle mass, temperature, and sometimes an MRI. He's recently started taking follistatin, a gene therapy that has been shown to extend the lifespan of mice. He drinks a pre-workout smoothie and takes approximately 60 pills, then does around 30 different exercises over the course of one hour, including using a high-frequency electromagnetic stimulation device that, he says, enables him to do 20,000 sit-ups in 30 minutes. By 9:00 AM Johnson has already eaten breakfast and lunch—steamed vegetables topped with undutched dark chocolate and hemp seeds, plus the aforementioned pudding with a quarter of a Brazil nut— and put himself through various treatments. Dinner, some type of salad, often with sweet potatoes or beets, usually takes place at 11:00 AM, followed by approximately 40 more pills. Bedtime is at 8:30 PM—which means he is typically fasting from around noon to 4:30 AM. On social media, he posts about his perfect sleep, his bone mineral density, his liver's "perfect enzymes," and his "urine flow rate," and periodically releases data that purports to show that his organs are younger than his chronological age, which is 46.[1]

This is the very cutting edge of transhumanism at work, though, if you ask me, he went too far with the chocolate on the steamed veggies. Johnson has received massive press coverage this past year.[2, 3, 4, 5, 6, 7] So why are we so captivated by this ambitious experiment? Firstly, it's like watching real-life science fiction unfold. We're witnessing the future of healthcare, a future in which aging might become just another wrinkle to be smoothed out. Secondly, Johnson's transparency is as magnetic as his ambition. He documents his journey meticulously, allowing us a peek behind the curtain of this grand biohacking experiment. It's a captivating blend of science, suspense, and a touch of forbidden knowledge.

But the real allure, the ultimate aphrodisiac in this biohacking odyssey, is the potential democratization of antiaging interventions. Imagine a world where the benefits of Johnson's discoveries aren't limited to the privileged few. We, the architects of the future, could all potentially extend our prime, granting us more time to make our mark, to leave the world breathless, and to come up with moonshot moments at unparalleled speeds.

The Rise of Transhumanism

We've already traversed the exciting landscapes of self-improvement for moonshots—from the tech-driven efficiency hacks to the mind-bending potential of psychedelics. But the journey doesn't end there. Now I want to dive more fully into the world of transhumanism, a revolutionary movement that takes human enhancement to a whole new level. We're not just talking about optimizing our performance within the confines of our current biology. We're venturing into a realm where technology itself becomes a catalyst for pushing the boundaries of what it means to be human.

Transhumanism is a way of thinking and acting that pushes the boundaries of what it means to be human. It embraces technological advancements, particularly those in bioengineering, artificial intelligence, and nanotechnology, to enhance our physical, cognitive, and emotional capabilities. Transhumanists envision a future in which we can overcome limitations like disease, aging, and even death, ultimately transforming ourselves into something beyond our current biological form. This movement isn't just about individual augmentation, it's also about creating a future in which these advancements are accessible and beneficial to all.

Transhumanism fuels moonshot moments by fundamentally shifting our perceptions of what's possible and our physical capacities to achieve those dreams. If psychedelics expand our minds, transhumanism pushes the limits of our brains and bodies. The pursuit of radical human enhancement challenges age-old assumptions about our limitations. It sparks bold ambitions to extend lifespans dramatically, engineer superior intelligence, and build physical bodies that can withstand the limits of our cosmos. These bold goals act as catalysts, driving massive investments in research and development, accelerating breakthroughs, and redefining the timeline of scientific progress.

Transhumanism, as I'll show, is a topic fraught with genuine ethical problems but is a key we need to unlock new eras of moonshot clusters, so we can solve the problems of today and venture into the cosmos tomorrow. It's time to stop being afraid of fire and figure out how to cook with it.

Transhumanism has been steadily rising in recent years, igniting excitement about the potential to transcend our current limitations while also sparking ethical concerns that demand careful consideration. We require a nuanced discussion that considers the potential for transhumanism to not only enhance the human body to tackle new frontiers and challenges but also to bridge the gap between what we can currently achieve and a future in which limitations are not barriers but opportunities for growth.

According to philosopher Nick Bostrom, transhumanism advocates for the enhancement of the human condition through applied reason and technology.[8] It welcomes the use of science to eliminate aging, augment our intellectual, physical, and psychological capacities, and ultimately transcend our current biological limitations. Transhumanists believe technology is a powerful tool that, used responsibly and ethically, can propel humanity into a new phase of existence, unlocking unprecedented possibilities for our future.

The origins of transhumanism are often traced back to evolutionary biologist Julian Huxley, brother of famed novelist and psychedelics-advocate Aldous Huxley. In his 1957 essay, "*Transhumanism*," Julian argued that

> [T]he human species can, if it wishes, transcend itself—not just sporadically, an individual here in one way, an individual there in another way, but in its entirety, as humanity. We need a name for this new belief. Perhaps transhumanism will serve: man remaining man, but transcending himself, by realizing new possibilities of and for his human nature.[9]

Defining transhumanism is tricky precisely because it's a dynamic and evolving idea, sparking ongoing debate about its true scope and goals. For some, transhumanism is about radically transforming humanity into something unrecognizable—a future in which humans merge with machines and upload their consciousnesses, achieving a form of digital immortality. For others like myself, it's about incremental advancements—curing diseases, extending lifespans, and subtly enhancing our existing capabilities. This lack of consensus on what constitutes the "endpoint" of transhumanism makes it difficult to pin down. Further muddying the waters is the question of whether we're already on the transhumanist path,[10] using current technologies like prosthetics, psychedelics, and nootropics (substances purported to enhance cognitive function) to slightly push our boundaries. This makes defining transhumanism a philosophical puzzle—are we changing the essence of what it means to be human, or merely enhancing it? Have we already started, and where are we going?

It's worth examining the major loci of research and discussion within transhumanism: What kinds of tech are we talking about when I say transhumanism in the first place? I'll begin with what I believe to be the most central theme within transhumanism: the quest to extend lifespans and combat the aging process. This category, often referred to as life extension or antiaging, encompasses a diverse range of research areas. Scientists are exploring various strategies, including biohacking (controlling intake and bodily experience to reverse or delay aging), regenerative medicine (using stem cells to repair or replace damaged tissues), and genetic manipulation (identifying and manipulating genes associated with longevity). For example, the SENS Research Foundation, led by scientist Aubrey de Grey, focuses on identifying and repairing seven underlying causes of aging,* aiming to achieve "engineered negligible senescence," a state where aging is no longer the primary cause of death.[11]

Another significant field of transhumanism is morphological freedom, which emphasizes an individual's right and ability to modify their body as they see fit.[12] This category embraces technologies like prosthetics, which can restore functionality and mobility to individuals with disabilities. Advanced prosthetics, like bionic limbs equipped with sensors and AI, are blurring the lines between human and machine, offering unprecedented levels of control and functionality.

---

* The seven underlying causes of aging include intracellular waste; intercellular waste; nucleus mutations; mitochondrial mutations; stem cells loss; an increase in senescent cells; and an increase of intercellular protein links.

Additionally, advancements in genetic engineering and body modification technologies raise questions about the extent to which individuals can and should alter their physical characteristics.

Superintelligence, another key category of transhumanist thinking, explores the potential creation of artificial intelligence (AI) that surpasses human capabilities in various domains. This raises exciting possibilities for scientific advancement and problem-solving but also sparks concerns about the potential risks and ethical implications of such powerful technology. Some transhumanists, like philosopher Nick Bostrom, advocate for careful research and development of AI to ensure its alignment with human values and prevent potential existential risks.[13] To complicate matters, discussions around superintelligence often shift to the realm of whole brain emulation and the hypothetical process of uploading a human mind into a digital format, potentially enabling a form of digital immortality.[14] The possibility of a technological singularity, a hypothetical moment in time when AI surpasses human intelligence and triggers unpredictable changes in human civilization, is another key theme within this category.[15]

Space colonization also features prominently within some strands of transhumanism. Proponents believe that establishing permanent human settlements on other planets is crucial for ensuring the long-term survival and expansion of the human species.[16, 17] This category draws inspiration from figures like Elon Musk and organizations like Mars One, who are actively developing technologies and fostering public interest in space exploration and potential human colonization of other celestial bodies.

Finally, democratic transhumanism emphasizes the importance of equitable access to human enhancement technologies. Proponents argue that if and when these technologies become available, they should be distributed fairly and not exacerbate existing social inequalities. This category raises crucial questions about the affordability, accessibility, and ethical implications of distributing powerful human enhancement technologies across different populations and socioeconomic groups.

Many, though not all, transhumanists hold a general optimism toward the potential of emerging technologies to benefit humanity, a view known as extropy elevated within the transhumanism field by philosopher Max More.[18] I'd certainly place myself happily in this camp—I'm one of many entrepreneurs who share an extropic perspective and believe that advancements in fields like AI, biotechnology, and nanotechnology can be harnessed to solve major global challenges like disease, poverty, and environmental degradation. The techno-optimists like myself strongly support the idea of moonshot moments, where transhumanism will give us a chance to expand to new realms of existence and push our bodies past previously held to limits.

As a techno-optimist within the transhumanist movement, I can't help but tingle with excitement at the sheer scope of possibilities that stretch before us! Imagine a world where firefighters bravely plunge into smoke-filled danger with augmented headgear, revealing the scene through the smoke and issuing warnings of impending dangers. Or envision future terraformers on Mars, their bodies genetically tailored to thrive in a low-oxygen environment, pushing the boundaries of human existence on a distant planet. We stand on the threshold of a future where technology seamlessly integrates with our biology, unlocking a potential we can barely comprehend. This is the thrilling promise of extropic transhumanism—a future in which humans transcend limitations and the unimaginable becomes our remarkable reality.

But even as an optimist, I believe that we have to look at the serious ethical considerations within transhumanism, to be sure we are embracing a future on the right terms. Just as we discussed technomoral virtues[19] in relation to the development of AI or synthetic biology, so too must we consider our values and virtues when considering transhumanism. Considering that transhumanism calls into question the very notion of being human, it's worth exploring some of the most pressing questions in the field.

"Fixing Us": The Ethics of Transhumanism

As we delve into the ethical labyrinth surrounding transhumanism, the first critical hurdle we encounter is the question of informed consent and bodily autonomy. With the potential for ever-more intricate interventions that alter our bodies and minds at a fundamental level, ensuring individuals make informed decisions becomes paramount. How can we guarantee that those considering transhumanist procedures fully grasp the potential ramifications, both positive and negative?

We can't even agree if lifesaving vaccinations are good for our children, so "informing the public" is not a sufficient benchmark for informed decisions around transhuman modifications. Additionally, the concept of bodily autonomy itself requires reevaluation. If we fundamentally modify our biology through genetic engineering, bionic implants, or brain-computer interfaces, does the concept of "my body" retain the same meaning? These are just the initial questions that arise within this complex ethical domain. As transhumanism pushes the boundaries of human potential, safeguarding individual autonomy and ensuring informed consent will be foundational pillars of a responsible and ethical future.

Transhumanist advancements, much like the development of AI, hold the potential to create a world far more equitable than our current one. Transhumanists envision a future in which cognitive enhancements bridge the learning gap, prosthetics become indistinguishable from natural limbs, and life expectancy transcends social determinants of health. Race, gender, and identity, as social constructs, lose their divisive nature. Transhumanist advancements could dismantle social injustices rooted in physical and cognitive limitations. But not all are convinced, transhumanism, like eugenics before it, can easily fall prey to existing biases. AI ethicists Timnit Gebru and Émile P. Torres warn of the dangers of what they call the "TESCREAL bundle": the collection of humanity-improvement philosophies including transhumanism, extropianism, singularitarianism, (modern) cosmism, rationalism, effective altruism, and long-termism.[20] They argue that all of these philosophies are derived from first wave eugenics, and that it is not possible to utilize philosophies to embrace equality when those philosophies originate in a mindset of racial superiority. It is important to take such criticisms seriously, and to reject any racially or culturally superior worldviews within the transhumanist movement.

The next realm demanding scrutiny is the potential for militarization of these advancements. Transhumanist technologies hold the potential to revolutionize warfare, with possibilities ranging from the creation of enhanced "super soldiers" with unparalleled physical and cognitive capabilities to the development of autonomous weapons systems capable of independent decision-making. The specter of such technologies raises a chilling question: Could the pursuit of transhumanist military dominance trigger a new and terrifying arms race? Considering our discussion of the concerns around arms races in the realms of AI and synthetic biology broadly, it would not be surprising to see a race in transhumanist genetic or robotic modifications on military personnel. The ethical responsibility to prevent such a scenario necessitates international dialogue and collaborative efforts to establish clear guidelines and regulations. Can we harness the potential of transhumanism to enhance human security without unleashing a wave of destructive weaponry? Furthermore, the very notion of a "super soldier" necessitates a moral evaluation. Would enhanced soldiers be viewed differently in society? Would the psychological burdens of war be amplified by heightened physical and cognitive abilities? These are just some of the weighty ethical considerations that must be addressed to ensure transhumanism empowers peace, not war.

The ethical landscape of transhumanism extends beyond individual considerations and military applications. Equally concerning is the potential for hacking and security breaches. As brain-computer interfaces and bionic implants become commonplace, directly interfacing with our neural networks, vulnerabilities will emerge. Malicious actors targeting these augmented systems could manipulate our perceptions, control our prosthetics, or even steal sensitive data. Robust security measures, constantly evolving alongside these technologies, are imperative safeguards. The very definition of privacy takes on a new dimension. If our thoughts and memories are increasingly intertwined with digital systems, ensuring individual privacy and preventing unauthorized access to this sensitive information is paramount. International cooperation on cyber-security protocols and clear ethical frameworks will be crucial to mitigating these risks. Building trust in a transhumanist future hinges on our ability to harness the power of these technologies while simultaneously protecting individuals from manipulation. To me, these are actually some of the most pressing practical concerns for transhumanism, and part of the reason that I argue later on that I don't think most people will embrace being "uploaded" or even having a direct connection to a digital world: the concerns about security and privacy will lead to slow adoption of many mental technologies, and perhaps rightly so.

The ethical considerations surrounding transhumanism extend further to the very definition of disability, exceptionalism, and uniqueness. Most agree on the positive applications of these interventions in restoring lost limbs or combating illness. For example, I think everyone has someone in their life who has battled cancer, maybe winning, maybe losing. And I think we all can agree that if synthetic biology and the application of transhumanist principles meant we could cure any cancer—or even prevent it—then we'd be on board. Similarly, what about children born deaf getting advanced cochlear implants and hearing their mothers sing? Yes, everyone's ready to get on that train.

But a crucial question arises: Where do we draw the line between a disadvantage and a unique human variation? In our quest to end illness, we have to ask what we lose. Sure, if we could fix a gene in a fetus with cystic fibrosis, so that neither parent nor child experiences that pain and heartbreak, it would benefit us all to put an end to the disease. But what about Down Syndrome, a disorder that still allows many with it to experience long, fulfilling, impactful, loving lives? Do we lose something within the larger human community by choosing to erase "abnormalities" in the genetic code?

Will we lose something when transhumanists find a "solution" to the suite of experiences we classify as neurodivergence, such as autism, OCD, ADHD, and more? What if we limit neurodiversity among our species? Ah, see, now we've lost our Salvadore Dalís, Steve Jobs, Issac Newtons, Michelangelos; we've lost our Darwins and Teslas and Einsteins and suddenly those moonshot moments seem to slip further from our collective grasp. If we had a transhumanist solution to ADHD, would I have ever written this book? Perhaps, but I would not have been so widely read nor so ready to make cross-disciplinary associations. Moonshot lost.

This is why transhumanism's potential to erase perceived imperfections presents a double-edged sword. Could the pursuit of a standardized, enhanced human form lead to a devaluation of human diversity and a pressure to conform? Conformity is the very antithesis of moonshot moments—no, it is rejection of the status quo that has fueled so many moonshot clusters. We must be wary of transhumanist approaches that seek to standardize our mental processes and capabilities, so that we do not stunt our ability as a species to reach for the stars.

Indeed, we may even find ourselves asking ethical questions about inducing abnormalities: If we could give you a pill that allowed for instantaneous mathematical computational abilities but caused severe behavioral problems, how many parents of aspiring astronauts would let their children take it? Would we sell it in vending machines at MIT and hire more campus counselors? Would it be popped at Silicon Valley parties like psilocybin is today?

Obviously, transhumanism's promise of curing diseases and enhancing abilities is thrilling. But these are the very questions that will establish our techno-moral virtues: Where do we draw the line between fixing problems and erasing what makes us unique? Transhumanism shouldn't be about creating a uniform humanity; it should celebrate diversity and fuel the kind of innovation that pushes boundaries. Would a world without the unique perspectives of those with disabilities, or the groundbreaking ideas that often come from those who think differently, be a world poised for progress? Transhumanism shouldn't strive for conformity; it should celebrate the richness of human variation to push us toward moonshot moments. As we push the boundaries of what it means to be human, we must be careful not to standardize our minds and bodies in pursuit of an idealized norm. After all, history teaches us that breakthroughs often come from those who see the world differently.

There is still one last ethical issue that concerns me, and that is most worthy of extensive discussion: the economic or class divide that inevitably comes with cutting edge transhumanist technology, and the inevitable exacerbation of existing economic inequalities that transhumanism will bring. Simply put: Upgrading tech is never cheap, and first adopters are usually the wealthy. What happens when the tech being upgraded is us, and, to put it cruelly, the poor don't get invited to the immortality party?

The dream of transhumanism, a future in which human potential is radically enhanced, crumbles without equitable access. If life-extending technologies and cognitive boosters become the privilege of the wealthy, a deepened societal divide is guaranteed. This isn't just morally repugnant, it stifles innovation. A world where only a select few benefit from these advancements squanders the potential for a richer tapestry of ideas and a broader range of minds tackling humanity's challenges. Furthermore, hoarding these advancements risks breeding resentment and potentially even conflict. Withholding lifesaving technologies in a world grappling with resource scarcity could lead to desperate measures. Transhumanism, for all its potential, can only flourish if it bridges divides, not widens them. As I'll discuss further on, international collaboration, innovative funding models, and ethical frameworks prioritizing accessibility are paramount. Only then can we ensure everyone, regardless of economic background, has the chance to participate in this new chapter of human existence.

The promise of transhumanism hinges on ensuring equitable access. We already see glimpses of the potential divide in our current world. The quality of prosthetic limbs often reflects the wearer's socioeconomic background. Nootropics and microdosing are trending in some circles but remain out of reach for many due to cost. Even basic medications like insulin highlight disparities in access to healthcare. The line between essential treatments and transhumanist enhancements may blur in the future. Already, rejuvenating creams, pills, and therapies have become yet another marker of wealth. To prevent a future segregated by access to these advancements, we must begin now to lay the groundwork for equitable distribution and ethical frameworks that prioritize the collective good over profit margins.

For now, however, much of the most cutting edge and experimental efforts in longevity and rejuvenation in particular are being funded and practiced by some of the world's wealthiest. The practice, today colloquially known as bio-hacking due to its close ties to Silicon Valley's elites, is far larger than just the psychedelics/microdosing movement, and is fraught with ethical and practical considerations of its own. The biohacking movement transcends celebrity fads and snake oil—it's a grassroots exploration of pushing our physical and cognitive limits in pursuit of a future filled with groundbreaking achievements.

Breathing Better: Biohacking for the Masses and Elites

The quest for peak human performance has taken on a new form in recent years: biohacking. This citizen science movement, particularly popular among well-heeled individuals in the United States and Europe, involves taking matters of health and well-being into their own hands. Biohackers experiment with a wide range of techniques, from optimizing sleep and diet with wearable trackers and personalized meal plans to exploring the potential of nootropics and even using cold exposure therapy to boost energy levels. This DIY approach to health optimization blurs the lines between traditional medicine and self-experimentation, raising both excitement and ethical concerns about the future of human potential. What was once fringe is now mainstream: experts predict the biohacking market will reach $63 billion U.S. dollars by 2028.[21, 22]

The seeds of biohacking were sown as early as the 1960s, when the counterculture embraced self-experimentation and challenged traditional notions of health and wellness. However, the term *biohacking* itself wasn't coined until the late 2000s, emerging from the San Francisco Bay Area's thriving DIY (do-it-yourself) biology and maker communities.[23] These early biohackers conducted simple experiments in makeshift labs, often focusing on understanding and manipulating basic biological processes. The rise of affordable technology like wearable health trackers and bioprinting tools further fueled the movement, allowing individuals to monitor their bodies and experiment with interventions in a more accessible way. Social media platforms also played a crucial role, fostering a sense of community and allowing biohackers to share experiences, protocols, and results.[24] Today, biohacking has evolved into a multifaceted movement encompassing citizen scientists, entrepreneurs, self-quantifiers, and even some medical professionals, all united by a desire to push the boundaries of human health and performance.

The explosion of wearable fitness trackers has propelled biohacking into the mainstream.[25] These devices, once limited to niche communities, have become ubiquitous, transforming biohacking from a fringe practice to an everyday activity. By monitoring heart rate, sleep patterns, and activity levels, these wearables empower individuals to become active participants in their own health journeys. The data collected allows users to not only track progress toward fitness goals but also experiment with different lifestyle choices and observe their impact on personal metrics.

In particular, the gamified approach to self-quantification, fueled by readily available data and user-friendly apps, has ignited widespread interest in biohacking, making it accessible to a broader audience beyond the realm of tech-savvy enthusiasts. As wearable technology continues to evolve, incorporating features like blood glucose monitoring and stress tracking, the boundaries between traditional healthcare and personalized biohacking are likely to continue blurring, shaping the future of health and wellness for millions.

Similarly, sleep optimization has emerged as a cornerstone of the biohacking movement, with roots tracing back to the early days of wearable technology. Pioneering companies like Fitbit and Jawbone popularized sleep tracking devices, empowering individuals to monitor their sleep patterns and identify areas for improvement. This data-driven approach resonated with biohackers seeking to quantify and optimize every aspect of their health. Today, sleep optimization is championed by a diverse range of individuals, from tech entrepreneurs and health influencers to professional athletes seeking to maximize their performance. Leading figures like Matthew Walker, author of *Why We Sleep*,[26] and Dr. Michael Breus, known as the Sleep Doctor,[27] advocate for the importance of sleep hygiene and offer practical strategies for optimizing sleep quality. Light therapy lamps, designed to mimic natural sunlight and regulate circadian rhythms, are another popular choice. Additionally, a wide range of sleep-promoting supplements like melatonin, magnesium, and L-theanine are available, though their effectiveness and safety vary.

Along the same vein, light hacking, the practice of manipulating light exposure to optimize health and well-being, is gaining traction as research provides more insight into the intricate relationship between light and our internal biological clock, the circadian rhythm. Today, light hacking finds itself championed by a diverse group, including researchers, sleep specialists, and even professional athletes seeking to optimize their sleep and performance.[28] The market reflects this rising interest with a surge in light hacking products. Light therapy lamps, mimicking natural sunlight and offering adjustable intensity and color temperature, have become a mainstay. Additionally, blue light-blocking glasses, designed to filter out the blue light emitted by electronic devices and potentially disrupting sleep patterns, are gaining popularity. Smartphone apps offering personalized light exposure schedules and reminders further empower individuals to take charge of their light environments.

We also see biohacking related to other sensory experiences and our health. For example, the surge of exercise hacking owes much to the desire for efficient and impactful workouts in a time-crunched world. While the precise origins are difficult to pinpoint, the philosophy aligns with earlier fitness trends like "minimum effective dose" training, which emphasizes maximizing results with minimal time investment. This concept resonated with individuals seeking to optimize their fitness routines. Today, exercise hacking finds strong proponents in various spheres. Tim Ferriss, author of *The 4-Hour Workweek*,[29] is a well-known advocate of exercise hacking, popularizing high-intensity interval training (HIIT) for its time-efficiency and potential performance benefits.

The roots of dietary biohacking can be traced back to the early days of alternative medicine and self-experimentation. Pioneering figures like Herbert Shelton,[30] a proponent of fasting for health, and Robert Atkins,[31] who popularized the low-carbohydrate ketogenic diet, laid the groundwork for this approach—a move away from "eat the food pyramid" to eating foods to specifically target a desired bodily experience. However, the actual term "dietary biohacking" gained traction in the late 2000s, fueled by the rise of citizen science and the growing interest in optimizing health through nutrition.

Today, many consider Dave Asprey to be one of the OGs of biohacking, with his *Bulletproof Diet*[32] emphasizing high-fat, low-carb meals and intermittent fasting. Similarly, Dr. Mark Hyman,[33] advocating for a personalized, "functional medicine" approach to nutrition, joins Asprey as the most prominent voices in the biohacking community. Many have jumped on the dietary biohacking band-wagon, leading to a smörgåsbord of products and services catering to dietary biohackers. Meal subscription services specializing in ketogenic or paleo diets, apps offering personalized fasting schedules and nutritional tracking, and even genetic testing kits claiming to provide insights into optimal dietary choices are just a few examples. Additionally, various supplements, ranging from ketone salts claimed to enhance cognitive function to fiber supplements promoting gut health, cater to the desire for optimized nutrition.

Beyond just food, the use of nootropics and supplements to enhance cognitive function has a long history within biohacking. Today, the use of nootropics and supplements to "hack" cognitive function finds proponents among students, entrepreneurs, and even professional athletes seeking an edge. Figures like Ben Greenfield[34] and Dave Asprey advocate for various nootropics like caffeine, L-theanine, and Lion's Mane mushroom, often emphasizing anecdotal experiences and personal experimentation. Companies like Thesis[35] seek to make nootropics accessible to middle-class, white-collar workers, and students.

The biohacking world has become increasingly fascinated with peptides, short chains of amino acids that act like messengers within the body. These tiny powerhouses can influence everything from metabolism to wound healing, making them attractive targets for biohackers seeking to optimize their health. However, peptides are a double-edged sword. Their effectiveness is often backed by limited research, and the lack of regulation raises concerns about purity, dosage, and long-term side effects. This controversy has biohackers navigating a murky landscape, often relying on anecdotal evidence and self-experimentation.

Ironically, the trend toward peptide use might be hiding in plain sight. The recent surge in popularity of Semaglutide (brand names include Ozempic and Wegovy), a GLP-1 receptor agonist, exemplifies this. While marketed as a diabetes medication, Semaglutide's weight-loss benefits have propelled it into the mainstream. What many users might not realize is that Semaglutide mimics the action of a naturally occurring peptide, GLP-1.[36] This mainstream adoption of a drug that essentially harnesses the power of a peptide hints at a potential shift in attitudes. Biohackers might be paving the way for a future in which peptide-based therapies, and perhaps other biohacking tools beyond simple diet and exercise, become more widely accepted and integrated into traditional healthcare.

While many of the biohacking practices I've described are accessible to a wide range of individuals, a growing niche caters specifically to the wealthy, venturing into the realm of extreme DIY biohacking. This exclusive world revolves around personalized medicine, cutting-edge technology, and luxury experiences, often blurring the lines between legitimate health optimization and experimental procedures.

One example is the emergence of personalized medicine companies offering comprehensive genetic testing, microbiome analysis, and advanced blood tests, all aiming to provide a personalized road map to health and longevity. These analyses, often costing tens of thousands of dollars, can be used to inform personalized diets, supplement regimens, and even potential interventions based on individual genetic predispositions.

Cold exposure therapy, the practice of intentionally subjecting oneself to cold temperatures, has roots in ancient cultures like the Romans and Greeks who utilized cold baths for medicinal purposes. However, its modern resurgence within biohacking can be traced back to the work of Wim Hof, a Dutch extreme athlete known for his ability to withstand extreme cold. Hof popularized the Wim Hof Method,[37] which combines specific breathing exercises and cold exposure to improve mental control, physical resilience, and overall well-being.

Biohackers like Ben Greenfield[38] and Dr. Rhonda Patrick[39] go further, exploring the potential of cold therapy for managing inflammation, boosting energy levels, and even enhancing cognitive function. The market reflects this rising interest with a surge in cold exposure products and services. Cryotherapy chambers, offering whole-body exposure to extremely cold temperatures, have become available at specialized wellness centers. Additionally, at-home cold therapy options like portable ice baths and cold shower attachments are gaining popularity. Athletes and Silicon Valley elites both embrace cold therapy; Lewis Hamilton and Jack Dorsey regularly feel the freeze.[40] Biohackers may also experiment with DIY methods like cold plunges in natural bodies of water, emphasizing the accessibility of this practice.

Sensory deprivation tanks, which remove all external stimuli, allow for the ultimate immersive meditation experience. Also known as isolation tanks or float tanks, they are soundproof, lightless pods filled with Epsom salt water. By eliminating external stimuli like sight, sound, and temperature variations, these chambers aim to induce a state of deep relaxation and sensory isolation. Meditating and resting in sensory deprivation tanks is expected to reduce muscle tension, improve circulation, and relieve pain. Similarly, while infrared light therapy has become a mainstream biohacking tool utilized in home saunas and muscle recovery devices, the world of extreme DIY biohacking offers high-powered and often custom-built infrared (IR) emitters.

Other examples of popular but expensive biohacking practices favored by the affluent include intravenous (IV) vitamin therapy, involving infusions of vitamins, minerals, and other nutrients directly into the bloodstream. Alternatively, another approach to high-priced healing is Hyperbaric Oxygen Therapy (HBOT), where individuals enter pressurized chambers and breathe pure oxygen. HBOT, though sometimes used in hospitals for specific conditions, has become a biohacking favorite among the affluent due to its potential for accelerating healing, reducing inflammation, and even promoting antiaging effects. However, this luxury doesn't come cheap—a single HBOT session can cost thousands of dollars, making it an exclusive playground for those with deep pockets.

Another arena for the ultra-wealthy is the world of biohacking retreats and immersive experiences. These exclusive retreats combine cutting-edge technology like sensory deprivation tanks and infrared beds, designed to promote deep relaxation, healing, and heightened awareness. Such retreats give visitors access to biohacking experts, personalized therapies, and even experimental treatments. Biohacking retreats, with costs reaching into the six figures, cater to a specific clientele seeking a transformative and rejuvenating experience.

Many of these more extreme forms of biohacking today are advocated by popular, albeit extreme, figures in the biohacking movement. For example, Dave Asprey, Tim Ferriss, and Ben Greenfield all consider themselves as prominent figures in the early days of biohacking, shaping its public perception and attracting widespread interest. Aubrey de Grey occupies a distinct space within the biohacking landscape, known for his advocacy of radical life extension methods.[41] De Grey, a gerontologist and cofounder of the SENS Research Foundation, proposes strategies like mitochondrial replacement therapy, aimed at reversing aging at the cellular level.[42] While his ideas are still far from realization and raise substantial ethical concerns, he represents a segment of the biohacking community actively exploring synthetic biology's frontiers of antiaging research and intervention.

Not all figures in the biohacking movement have survived at the experimental edge. A cautionary example is Aaron Traywick, a biohacker who was more of a snake oil salesman than legitimate transhumanist. Traywick gained notoriety within the biohacking community for his ill-fated attempt at DIY gene-editing.[43] He experimented with CRISPR-Cas9 technology, a powerful gene editing tool still under development, without proper training or medical oversight, with the promise of developing and sharing synthetic biology solutions to AIDS and herpes. He was best known for publicly injecting himself with his creations, doing significant damage to public perceptions of transhumanism.[44] This risky self-experimentation tragically resulted in his death in 2018. Traywick's story serves as a stark reminder of the importance of responsible biohacking practices. It highlights the dangers of unregulated and poorly understood interventions, especially when dealing with complex technologies like gene editing.

Ultimately, Silicon Valley thrives on innovation and optimization. It's no surprise, then, that the tech world is particularly drawn to biohacking. Here, biohacking isn't just about kale smoothies—it's about leveraging cutting-edge technology and scientific research to push the boundaries of human potential. Tech titans see themselves as constantly optimizing their software, and biohacking offers the tantalizing possibility of optimizing their own hardware—their bodies and minds—for peak performance, creativity, and longevity. It's the ultimate hack for the ultimate hackers, a way to extend their prime and keep innovating at the forefront of a rapidly changing world.

Shaking It Up: My Experiences with Biohacking

My fascination with biohacking started young, long before it had a fancy name. As a teenager, I wasn't just interested in looking good, I craved a deep understanding of how my body functioned. I began with the fundamentals: fine-tuning my nutrition. This wasn't about fad diets; it was about meticulously tracking my macros and experimenting with different food combinations to see what fueled my energy levels and athletic performance.

To optimize my workouts, my parents enlisted the help of a knowledgeable personal trainer. Together, we crafted personalized routines that pushed me to new limits. Biofeedback became a core aspect of this journey. I remember the thrill of getting regular DEXA scans, a precise measurement of body composition and body fat that went far beyond the limitations of a standard weight scale. It allowed us to track my progress objectively, ensuring my efforts translated into real results. To further refine my approach, I even built a shared Instagram account with my trainer. Posting pictures of my meals to ensure their balance provided a layer of accountability and fostered a collaborative spirit—an essential ingredient in any successful biohacking adventure.

College marked a turning point in my biohacking journey. Fueled by newfound independence and a thirst for deeper knowledge, I ventured beyond the realm of food and exercise. Blood tests became my new frontier, a window into the inner workings of my body. These tests, a far cry from the finger pricks of childhood, provided a wealth of information—cholesterol levels, hormone balances, and even markers of inflammation. Armed with this data, I could make informed decisions about my health. Supplementation became a strategic tool, not a random grab bag of vitamins. My doctor, trainer, and I meticulously researched and incorporated targeted nutrients to improve my cholesterol profile, bolster my immune system, and enhance focus during those late-night study sessions.

Most recently, I've launched into the fascinating world of genetic testing. Unlike the basic DNA ancestry kits, these tests offered a deeper look into my genetic code. They revealed not only predispositions to certain body compositions but also explained why certain processes, like muscle building or fat metabolism, might be more challenging for me. This personalized road map has empowered me to tailor my biohacking approach with even greater precision.

My biohacking journey wasn't solely driven by optimization; it was also fueled by a persistent enemy—pain. A childhood horse-riding accident left a legacy of damage in my lower spine, manifesting in a dull ache that could escalate into debilitating discomfort. Initially, I treated it like a stubborn foe to be wrestled into submission—through stretching routines and uncomfortable back braces. While these provided some relief, they weren't a long-term solution. My exploration of alternative therapies led me to the world of yoga and chiropractic care. These practices offered a holistic approach, addressing not just the physical symptoms but also the underlying stress and tension that could exacerbate the pain.

More recently, I've ventured into the realm of integrative medicine. Dry needling, a technique involving thin needles inserted into specific trigger points, has become a powerful ally, offering immediate pain relief. It embodies the spirit of biohacking perfectly—combining traditional healing practices with modern science. However, the most cutting-edge intervention in my arsenal is regular stem-cell based medicine, administered intravenously. While still a relatively new therapy, it holds immense promise for repairing my degenerated disc and achieving a level of pain management that once seemed impossible.

My biohacking journey also includes my epic struggle against dreadful allergies, a lifelong companion since childhood. The scratchy eyes, runny nose, and constant sniffles were all too familiar. As a kid, I underwent a battery of allergy tests, the culprit being identified as the ubiquitous oak and its pollen accomplices. Over the years, nonprescription medication became a mainstay, offering relief but never a lasting solution. Most importantly, I could not stand the drowsiness and clouded thinking that most allergy medications caused—it was a side effect that weakened all my thinking, working, and writing.

However, biohacking offered a glimmer of hope beyond allergy symptom management. I discovered immunotherapy, a fascinating approach that promised to address the root cause of my allergies. This cutting-edge treatment involves being injected with microdoses of the very substances I'm allergic to. The idea is to gradually train my immune system to recognize these allergens as harmless, essentially desensitizing it. The success story of my own father, who achieved complete allergy freedom within three years of undergoing this treatment, fueled my optimism. While the journey might be long, the prospect of a life free from allergy symptoms, achieved through a proactive approach, makes every microdose worth it.

The cornerstone of my biohacking approach today is a powerful yet unobtrusive tool: the Oura Ring. This sleek, sleep-tracking device has become an indispensable extension of myself. It goes far beyond counting sheep, though. The Oura Ring monitors a symphony of internal processes—heart rate variability, sleep stages, body temperature, and even recovery scores. This real-time data stream allows me to see how my body is functioning throughout the day, from the moment I wake up to the moment I drift off to sleep.

More importantly, the Oura Ring acts as a feedback loop for my other biohacking efforts. Did that new post-workout supplement actually improve my recovery score? Did the switch to a high-fiber breakfast lead to more stable blood sugar levels? The ring provides the answers, allowing me to refine my approach with laser focus. This data-driven approach extends beyond the ring itself. I prioritize a good night's sleep, tracking my sleep cycles and making adjustments to my routine as needed. My mornings often begin a workout routine I've faithfully maintained for almost eight years, followed by a ten minute sweat session in a 110-degree Fahrenheit steam room. Consistency is key, and the Oura Ring helps me stay accountable.

Beyond exercise, I've incorporated a variety of targeted interventions. My digestion gets a boost with a daily dose of a specific prebiotic supplement. My day is always started off with a Flintstones-vitamin flavored smoothie, the recipe for which I've meticulously tweaked over the years to ensure its flavor is edible and is packed with biohacking benefits. Cellular recovery is aided by a PEMF (pulsed electromagnetic field) device. And for those times when I need a deeper recharge, I utilize an infrared light therapy sleeping bag. To address potential nutrient deficiencies, I get regular intravenous infusions packed with essential vitamins and minerals. This biohacking journey may seem complex, but the core principle is simple: listen to your body and respond accordingly. The Oura Ring empowers me to do just that, transforming biohacking from a guessing game into a strategic pursuit of optimal health and well-being.

I acknowledge a critical aspect of my biohacking journey—I have the wealth to access many treatments and therapies that would be out of reach for many. While I'm fortunate to have access to a wider range of treatments and therapies, the good news is that the landscape is shifting. Costs are steadily decreasing for many biohacking tools, and I wholeheartedly believe this trend should continue. I'll make my argument shortly that biohacking shouldn't be a privilege for the few; it should empower everyone to take charge of their well-being.

But part of the beauty of biohacking lies in its scalability. You don't need a high-tech arsenal to begin. Starting with a simple health tracker like the Oura Ring, or even a more affordable alternative, can be your first step. Focusing on foundational elements—nutrition, exercise, sleep hygiene, and stress management strategies—can be a powerful biohacking tool in itself. These practices are cost-effective and universally applicable, laying the groundwork for a healthier, more vibrant you.

An Argument for a Democratic Approach to Transhumanism

For the West to maintain its position at the forefront of transhumanist research and innovation, we need a multifaceted strategy that goes beyond the pockets of wealthy biohackers currently leading the charge. A truly transformative movement demands a convergence of public acceptance, a rigorous research program spanning medical and technological fields, and a seamless collaboration between academia and industry. We need to embrace a democratic, egalitarian view of shared progress and shared transhumanism—the win-win route to more moonshot moments. Biohacking by a wealthy caste of elites is not enough for species-wide wild thinking.

Instead, we require hyper-cooperativity across academia, industry, and government to develop a coordinated and intentional transhumanist program. This concerted effort will foster the development of crucial safeguards and ethical frameworks to guide the expansion of transhumanist technologies and practices. By embracing inclusivity and democratic principles, we can work toward a future in which the benefits of transhumanism are accessible to all, shaping a society in which enhancements aren't dictated by wealth but distributed thoughtfully and equitably to better humanity as a whole.

So what would such a project look like, and how do we get the public on board? To cultivate a truly democratic transhumanist movement, it's imperative we shift the spotlight away from wealthy, fringe biohackers who often dominate media attention. Instead, we need to center our narratives around the demographic with the most immediate potential to benefit from the fruits of transhumanist research: individuals suffering from illnesses or debilitating conditions. Highlighting the transformative impact of prosthetics for amputees, the promise of genetic therapies for those battling chronic diseases, and the potential of brain-computer interfaces for neurological disabilities will resonate deeply with the public. We need to see collaborative efforts between universities and businesses, as well as more government funding to support such research, and then there also needs to be an active effort by the transhumanist community to bring attention to these collaborative efforts. This shift will humanize transhumanism, demonstrating its power to improve lives and alleviate suffering, shaping a broader perception of the movement as a force for good, not simply a playground for the privileged few.

And then that's where our immediate research should be going. To propel ourselves onto a truly transformative trajectory, we must boldly invest in highly innovative transhumanist research, including the field of psychedelics. This aligns seamlessly with the transhumanist goal of enhancing human experience and well-being. Alongside this, we must prioritize advancements in genetic engineering like CRISPR technologies, which have the potential to unlock cures for devastating diseases. Investment in cutting-edge robotics could lead to bionic limbs with unparalleled dexterity, restoring function to individuals with disabilities. Finally, fostering innovation in brain-computer interfaces like Neuralink promises breakthroughs that could assist individuals with paralysis or neurological conditions to communicate and interact with the world in novel and empowering ways. Venture capital firms should be backing small startups with fantastically new ideas. Such a multifaceted, forward-thinking approach to funding will pave the way for a future in which transhumanism truly serves humanity and will be a first step toward a larger public acceptance of transhumanism: where the money goes, the people will follow.

The key to unlocking a truly transformative future is to make sure there is no gatekeeping of such technologies. Have we figured out a gene we can switch on to prevent diabetes? Then we should make the procedure standard and free rather than going the route of prohibitively expensive insulin. Are we making prosthetics better and better? Then why do veterans even have to pay for them? In order for an ethical transhumanist future to occur, we must make sure that technologies that heal the ill are given to all regardless of wealth, and that advancements are shared indiscriminately.

It's a tall order, I know, but if transhumanism is a push to become a better version of ourselves physically, then perhaps we should strive to become better versions of ourselves morally as well.

Along with ensuring that medical advances are shared, we should also be funding companies that embrace bold solutions to difficult or dangerous jobs and positions—and I really don't just mean for soldiers and cops. We need to move beyond a paradigm of advanced robotics or genetic engineering as a weapon, and instead embrace practical applications for positions like firefighters, search and rescue teams, offshore miners, and space explorers. Again, the focus of transhumanism should be on the improvement of society as a whole, not the enhancement of individuals seeking a "get physically rich quick" scheme to gain a private path to immortality.[45] Nor should we focus on building better soldiers (though, inevitably, I fear we will, and as there may be other life in the universe, we likely should).

These society-oriented advances should be the focus of our research, before a drive by the wealthy to live forever. There will inevitably be the immortality focus of fringe experimenters, which is why transhumanism has always had a strain of punk-DIY mentality throughout discourse in the community. But advances in realms of restorative medicine and feature optimization will ultimately lead to advances in longevity studies and brain-computer interfaces. In other words, researching fields for the common good will certainly generate a lot of meaningful data for those focusing on antiaging and brain enhancements. There will be other moonshot moments within transhumanist research in the decades to come, propelling our understanding of our bodies forward: by sharing them cooperatively, and ensuring collaboration, we will continue to open a space for even more moonshot realizations and discoveries.

Chasing Unicorns in Texas

On a sunny summer day in September 2024, I found myself in Austin, Texas, sitting in an Uber headed toward a nondescript strip mall of office buildings. My destination was a shared laboratory space housing two fascinating startups: The Odin—dedicated to "making science and genetic engineering accessible"—and the Los Angeles Project, a new venture with the audacious goal of turning science fiction into reality. It wasn't what I'd imagined for their headquarters. The name "Los Angeles Project" conjured visions of sleek research campuses, futuristic facilities, or even something out of a comic book—like Xavier's School for Gifted Youngsters from X-Men—not a modest converted warehouse. Yet, within this shared space, cofounders Cathy Tie and Jo Zayner were hard at work on something extraordinary: bio-engineering a real-life unicorn, complete with its own mythical horn.

Cathy Tie, a rising star in the biotech industry, is a visionary entrepreneur with a passion for democratizing science. Her journey began at the University of Toronto, specializing in molecular genetics and rare genetic mutations. Recognizing the transformative potential of technology, Tie cofounded Ranomics, a genomics screening company, while still in her early twenties. Her innovative approach and entrepreneurial spirit caught the eye of tech mogul Peter Thiel, who selected her as a Thiel Fellow. This recognition propelled Tie's career, leading to the founding of Locke Bio, a telemedicine company aiming to streamline healthcare services, and the Los Angeles Project, a revolutionary gene-editing company. Her work at the intersection of technology and biology has earned her acclaim in Silicon Valley and beyond, solidifying her position as a leading figure in the biotech industry.

While Cathy embodied the entrepreneurial spirit of the team, cofounder Dr. Jo Zayner, a biohacking pioneer with a rebellious glint in her eyes, represented the scientific driving force. After earning a PhD in biochemistry from Berkeley, Jo cofounded The Odin, a biohacking lab famous for its DIY gene- editing kits, which now shares its research space with the Los Angeles Project. Their cutting-edge gene-splicing laboratory, where Zayner works with her team of biohackers each day, has become a place to turn childhood fantasies into reality.

Cathy and Jo led me on a tour of their lab, a vibrant hub of scientific innovation that felt as much like an art studio as a research facility. The atmosphere was electric, filled with a diverse and enthusiastic team working on a variety of projects. Unlike the often exclusive or rigid environments of academic labs or Silicon Valley startups, this space exuded inclusivity and creativity. Lab technicians, many of whom were self-taught under Jo Zayner's mentorship, collaborated seamlessly with seasoned researchers and fresh-faced interns, fostering an open exchange of ideas and mutual support. Jo's philosophy—that lab work and gene editing are as much about art and dreams as they are about science—was evident throughout the space. Scattered among the cutting-edge experiments were art projects and reminders of the beauty and limits of the natural world, infusing the lab with inspiration and a sense of wonder.

One particularly unforgettable moment was seeing a glowing axolotl in a tank—a gene-spliced marvel created by isolating fluorescent genes from a jellyfish and integrating them into the axolotl's DNA. This remarkable feat was a striking testament to their groundbreaking work and the imaginative spirit that drives it.

Following the lab tour, I sat down with Cathy and Jo for a lengthy interview to dig deeper into the driving force behind their transhumanist endeavors. Cathy and Jo envision a future where gene editing and CRISPR technology are as accessible as personal computers. "The way we see [biohacking] technology is that it can actually be even more powerful than nuclear weapons," Cathy explained, "but the current system stifles development . . . we kind of wanted to create our own box of what we can be and what the technology can be." Cathy not only saw their work as a modern manifestation of the Manhattan Project, but an opportunity to change how science could be used to rewrite our existence.

Sitting beside Cathy and dressed more like a punk rocker than a Berkeley-trained biofuturist, Jo Zayner offered an explanation of the project that was a bit more direct:

> This is the first time in the history of the fucking Earth, in the history of life on Earth, that we can change DNA; we can recode life, the DNA of living things. The first time! And people are like, "I'm gonna make weird lab mice and maybe a gene therapy to treat a disease." Good, great, great for you, but we can program self-replicating matter, right? That's what biology is, it's self-replicating matter. We can literally program that shit and that's all you wanna do? Don't you wanna create fucking Pokemon or unicorns or some shit? I mean, that's what I wanna create. Why aren't we trying to do that? So, I think that's what [the Los Angeles Project] all stems from: Cathy and I asking, "Why? Why aren't we trying to do that? Why aren't people trying to push these boundaries instead of just settling for boring shit?"

Jo's work at the Los Angeles Project sought to reset those boundaries. Forget the metaphoric 'unicorn' of billion-dollar Silicon Valley startups. The Los Angeles Project's mission was direct: to genetically engineer a unicorn. The proof of concept is designed to demonstrate the moving goal posts achievable through the revolutionary power of gene editing. While this endeavor continues to raise eyebrows and ethical concerns, it perfectly encapsulates the renegade spirit in the field of transhumanism, as well as the drive to cultivate moonshot moments. Cathy and Jo weren't afraid to court controversy in their pursuit of what they saw as progress.

Witnessing their dynamic in person was captivating. Cathy, the strategist, balanced Jo's maverick tendencies, channeling their scientific brilliance into a coherent vision. Together, they were a force to be reckoned with. They didn't just talk about disrupting the future; they see themselves as actively building it, brick by gene-edited brick.

Their efforts to create a unicorn are a proxy for their longer-term ambition: understanding and manipulating the human genome so that we can manage the challenges of our environment ahead. Ultimately, their vision is to unlock the potential of human evolution, enabling us to adapt to extreme environments and overcome physical limitations. Could we someday be able to breathe underwater, withstand greater atmospheric pressure, or gain immunity to UV damage to our skins? If humanity is to pursue a cosmic destiny of space exploration and settlement in our future, we will need ambitious transhumanist thinkers like Cathy Tie and Jo Zayner pushing the boundaries of scientific research today.

Whether or not they succeed in creating woodland wonders remains to be seen. But their audacity is undeniable, and they represent an extreme nexus of imagination and research that marks the far edge of transhumanism and Coming Wave technology. Cathy, Jo, and the entire Los Angeles Project team are leading a charge toward moonshot moments, driven by a potent mix of ambition, cooperation, accessibility, and a willingness to take risks. Their story is a testament to the power of human ingenuity, unshackled by convention and fueled by a desire to push the boundaries of what's possible. As I left the Los Angeles Project later that day, I couldn't help but feel a spark of their infectious optimism.

Maybe, just maybe, a unicorn isn't so far-fetched after all.

Giving the Dragon Wings: China's Transhumanist Movement

So far, our exploration of transhumanism has primarily centered on Western debates and examples. But it's crucial to realize that the race toward a transhumanist future is a global one. Just as China surges forward in the realm of artificial intelligence, it's also rapidly advancing in the fields of synthetic biology, nanotechnology, and robotics—all foundational technologies crucial to transhumanist goals. Within China, the path toward these advancements is marked by a different set of ethical considerations and regulatory frameworks than in the West. This distinction raises complex questions about the balance between innovation, individual rights, and societal implications as we navigate this emerging technological frontier. It also reminds us, again, that hypercooperation is important to ensure standards for all that ensure species-wide success. It's worth taking a moment to pause and look at just some of the transhumanist projects that are running in China.

While there has been a lot of buzz lately about Elon Musk's Neuralink and its potential, China has emerged as a major force in the race to develop and implement powerful brain-computer interfaces (BCIs). These cutting-edge technologies hold immense potential for revolutionizing various sectors, from healthcare and communication to education and entertainment. BCIs create a direct communication channel between the human brain and external devices, bypassing traditional communication pathways like muscles or speech. This allows for the transmission of information and control signals directly from the brain, opening doors to previously unimaginable possibilities.

Driven by an ambitious combination of state-backed initiatives and private sector innovation, China's BCI research is rapidly gaining momentum. The country boasts a vibrant ecosystem of research institutions, technology companies, and startups, all actively contributing to advancements in BCI technology. This intense focus positions China as a key player in shaping the future of this transformative technology.

One major area of focus is neurorehabilitation. Chinese researchers are developing BCIs to help individuals with paralysis regain control of their limbs or communicate more effectively. Studies explore BCIs for stroke recovery, aiming to restore lost motor functions and improve quality of life.[46] Beyond rehabilitation, there's also a growing interest in utilizing BCIs for potential cognitive enhancement. While still in its early stages, this research direction raises important ethical implications regarding the augmentation of human capabilities. China's unique regulatory environment, in which individual rights may sometimes be weighed against broader societal goals, adds another layer of complexity and concern to the development and deployment of BCI technology in the country.

China is also actively pushing the boundaries of longevity research through its focus on antiaging therapies and regenerative medicine. This pursuit stems from a deep cultural reverence for longevity and aligns with the broader transhumanist vision of extending lifespans and enhancing human health.

Numerous research initiatives are exploring various antiaging strategies. One notable example is the BGI Group, a leading genomics research company in China, which launched the "Healthy China 1 Million Genome Project" in 2018.[47] This ambitious project aims to sequence the whole genomes of one million volunteers, hoping to identify genetic markers associated with longevity and age-related diseases.

China also has a larger compendium of compounds to study in transhumanist research, with thousands of years of botanical learning. Several Chinese academic teams have been exploring the potential of naturally occurring compounds found in traditional Chinese medicine for antiaging applications.[48, 49]

China is also investing heavily in regenerative medicine, aiming to repair or replace damaged tissues and organs, potentially contributing to lifespan extension. One prominent example is the Guangzhou Regenerative Medicine and Health Institute, established in 2017.[50] The institute focuses on stem cell research and its applications in treating various diseases and potentially promoting tissue regeneration. Furthermore, companies like Beijing BGI Regenerative Medicine Technology Co., Ltd. are exploring the use of stem cell therapy for age-related conditions like osteoarthritis, demonstrating the translation of regenerative medicine research into potential clinical applications.[51]

Perhaps most concerning is China's rapid advancement in the field of genetic enhancement, where the government's loose regulatory regimes and history of pushing for extreme excellence makes an environment ripe for ethically gray genetic engineering. China's exploration of genetic enhancement and its ethical boundaries gained global attention in 2018 with the He Jiankui CRISPR experiment, a landmark event that sent shock waves through the scientific community and beyond.

He Jiankui, a biophysicist then working at the Southern University of Science and Technology in Shenzhen, announced the birth of the world's first gene-edited babies. He used CRISPR-Cas9 gene-editing technology to modify a gene called CCR5 in human embryos, attempting to confer resistance to HIV infection.[52] While the implications are exciting—being able to eradicate HIV through a treatment that would make future generations immune before birth—the research also raised serious implications for the field of eugenics.

He Jiankui's experiment was met with widespread condemnation and alarm within both the Chinese and global scientific communities. The ethical implications were enormous—he effectively ushered in an era of "designer babies," where the human genome could potentially be altered at will. The scientific community criticized the procedure for its risks and the lack of rigorous safety protocols,[53] arguing that the technology was not sufficiently developed for human germline editing. Additionally, there was no medical necessity to justify this controversial experiment as there are existing methods to prevent HIV transmission from parents to offspring.

This event ignited global debate on the ethical, social, and regulatory gaps surrounding human germline editing (editing your individual genes in a way that will be passed on to the next generation).[54, 55] It highlighted the need for clear international standards and a comprehensive framework to prevent rogue scientists from pursuing ethically questionable experiments. China itself subsequently tightened its regulations on gene-editing research, and He Jiankui was sentenced to three years in prison for his actions.[56] However, his experiment serves as a stark reminder of the potential for both groundbreaking advancements and ethical quagmires within the rapidly evolving field of bioengineering, particularly in the pursuit of transhumanist goals.

Staying Bodied: A Short-Term View of Humanity

While the complete transformation envisioned by some transhumanists may seem far off on the horizon, the near future holds exciting possibilities for the individual experience. In the next chapter, I'll consider a broader transhumanist vision that encompasses societal and technological shifts and space exploration. But I also want to close out this section of improving the self to create more moonshot moments with a discussion of what, short term, life might look like for the next two hundred years on earth. For now, we'll peek into a future not of radical disembodiment, but of an enhanced physicality.

I often imagine a world where individuals remain grounded in their physical selves, yet empowered by advances in regenerative medicine and cognitive augmentation. Let us contemplate the implications of a constantly renewing body and brain, exploring how these advancements can ensure the continuity of personal experience. And with such changes, we will definitely see social consequences, some of which we've already discussed in our journey through psychedelics, and some of which we'll expand on below.

The debate surrounding mind uploading and AI often gets bogged down in the philosophical quagmire of consciousness. Will we be creating brand-new consciousnesses in machines, or will we be able to transfer our own into them? As we change our own bodies and become less "physically human," will some form of consciousness remain? While these are certainly intriguing questions, I believe they might be ultimately unproductive. After all, we still don't have a universally agreed-upon definition of consciousness. It's a complex phenomenon, deeply intertwined with our physical brain and experiences, and trying to pin down its essence feels like trying to grasp a wisp of smoke. Debating something we can't even definitively define feels like chasing a meme, a fascinating but ultimately intangible concept.

Furthermore, the entire question hinges on the assumption that any AI or advanced humans we create will possess consciousness in a way similar to today's humans. This might not be the case at all! AI could evolve in ways completely unforeseen, developing its own unique form of sentience or intelligence that defies our current understanding. There's no reason to assume AI intelligence will look like ours, nor should we assume AIs will want to take human form as androids. If AI generates better models than "gender" or "self," should we ignore these and stick to our current formatting?

Additionally, the idea of extracting our minds from our bodies to exist purely in a digital realm strikes me as far-fetched and, frankly, undesirable, at least in the near future. Security is a pretty serious concern. So is corporality, or embodiment. Our bodies are an integral part of who we are, shaping our experiences and interactions with the world.

Just as I described humans' reluctance to let go of religions and existing beliefs in the face of new technology, I believe most humans will also express a reluctance to let go of their physical bodies to upload. Think of the primal, visceral joy of enjoying a delicious meal, the surge of endorphins and deep satisfaction after a challenging workout, or the electrifying spark of physical intimacy. Consider the awe-inspiring beauty of the natural world—the feeling of sunlight on your skin, the rush of a cool breeze, the symphony of birdsong. For me, I cannot imagine a world where I could not feel the sensation of riding a horse. Even the simple act of sharing a warm embrace with a loved one is inextricably linked to our embodiment. These experiences shape who we are and illustrate that so much of what it means to be human is fundamentally tied to the vessel that contains our consciousness.

This deep-rooted connection to our physical selves is precisely why I believe the near future will focus on technologies that enhance, rejuvenate, and repair our bodies and, particularly, our brains. While the human experience is deeply connected to our physical bodies, augmenting brain chemistry through science offers intriguing possibilities for enhancing our cognitive abilities. This can be achieved through various methods we've been discussing, including strategic nutrition, the exploration of nootropics, and even the potential of psychedelics. Unleashing our inner geniuses might be closer than you think. Optimizing brain chemistry through various tools offers exciting possibilities for experiencing moonshot moments of profound creativity and problem-solving.

Alongside this, advancements in brain-computer interfaces, genetic engineering, and potential implants hold the promise of pushing cognitive boundaries. These augmentations aim to improve memory, focus, creativity, and perhaps even empathy; they aim to unlock the latent potential of our minds while still grounding us squarely within our physical forms. The progress that Neuralink has already made with implants that help individuals interface with computers, to do everything from type commands to playing games,[57] makes me confident we will have well-functioning brain-computer interfaces in our lifetimes. I suspect we will see hyperspecialization as well: for example, I think we could agree that every astronaut should be given a mental implant that can do complex computational math on the fly, so that they can solve problems in real time even when they require complex equations.

In the grander vision I see for our future, mental implants capable of uploading and downloading information directly into the brain could revolutionize the very concept of learning, knowing, and remembering. Colloquially, we will plug in, sit for a few minutes, stand up, and say, "I know Kung Fu." These implants wouldn't simply be passive repositories of information; they could potentially work in tandem with our existing neural pathways, actively facilitating the absorption and integration of new knowledge and skills. An aspiring artist could, in theory, download not just the technical aspects of painting but also the years of experience and nuanced brushwork of a master.

Instead of spending years acquiring knowledge and honing skills, imagine the possibility of effortlessly uploading entire skill sets, bodies of knowledge, or even foreign languages tailored to one's individual mind. This wouldn't eliminate the need for creativity, critical thinking, and individual application, but it could drastically alter the landscape of learning and knowledge acquisition, democratizing access to expertise and potentially compressing the learning timelines for a vast array of human endeavors. This leap forward could unlock unlimited moonshot moments while profoundly redefining our educational systems and the way we interact with the vast, complex world around us.

The future I envision paints a picture of infinitely prolonged mortality, not absolute immortality. While advancements in medicine and body repair could significantly reduce the natural causes of death, accidents, unforeseen events, and perhaps even the choice of an individual could still lead to the end of life (essentially, engineered negligible senescence).[58] But with lifespans potentially stretching to centuries, the concept of death would transform. We've already discussed the implications for religion if the traditional concept of an afterlife fades away. Now, we must grapple with issues of overpopulation, resource allocation across generations with vastly different experiences, and the very meaning of existence in a world without the natural endpoint of death. As humanity embarks on this unprecedented journey, navigating these complex questions and forging a new social contract will be crucial to ensuring an equitable and fulfilling future for all.

For example, as Robin Auer points out, "on a very practical level, our whole legal system would be in shambles once immortality became a reality. Not only would the severity of certain crimes be altered dramatically, but we would also have to completely rethink the way punishment works."[59] Traditional inheritance laws, built on the inevitability of generational turnover, would need revision. Concepts like property ownership and wealth distribution across centuries would require new considerations. Criminal justice systems might grapple with the implications of potentially imprisoning individuals for lifetimes that could span centuries. Rehabilitation and restorative justice could take precedence over punitive measures. Additionally, the definition of personhood and the rights associated with it might need reevaluation in a world where artificial intelligence and potentially uploaded consciousness could exist alongside vastly extended human lifespans. Crafting legal frameworks to address these complexities will demand a global conversation, ensuring equitable application of laws across a society with an unprecedented lifespan and a diverse range of beings.

I truly believe that a transhumanist, extropic perspective, especially with a focus on extending lifespans and boosting our minds, is the key to unlocking a future brimming with potential. Sure, a world where death is a distant prospect might throw some curveballs—new social structures, new religions, new legalities we haven't even dreamed of. But these are challenges we can overcome together, through reason and hyper-cooperation. It's really not that hard to imagine it: regenerative medicine keeping us healthy for centuries, psychedelics and smart drugs sharpening our focus, and brain implants opening up entirely new pathways of learning. By essentially upgrading our biological hardware, we'd be giving ourselves the capacity to run more complex programs, to experience those incredible moonshot moments of innovation and problem-solving on a regular basis. This transhumanist vision isn't just about living longer; it's about living a life overflowing with creativity, pushing the boundaries of what it means to be human, and leaving a legacy that echoes not just for years, but for centuries to come.

And just how far can we go as a species? Earlier, when you chose your own future, I was able to show you some of the wildest, most exciting, and most terrifying possibilities for our future as a species. Many of these were highly speculative, even preposterous—designed to pique your curiosity about the inevitable wave of technologies and social changes that are coming. But now that we've explored just how possible it would be for us to make our societies and ourselves ripe for moonshot clusters, I want to discuss what I believe to be the logical progression of humanity and all our moonshot moments:

Let's finally talk about shooting to the moon and far, far past it.

PART IV

EMBRACING
OUR COSMIC DESTINY

Chapter Nine

RACING TO THE COSMOS

Thirty thousand years ago, early humans sat around their fires, eating cooked mammoth meat, telling stories, and singing songs together. Their eyes followed the smoke up into the sky, toward the little specks of light in the darkness. Did one of these ancient ancestors dream that someday, their greatest grandchildren would reach for those stars, not with hands outstretched, but in rockets and ships that even now we cannot fully imagine?

I truly believe that they did, in their own way. For millennia, humanity has gazed skyward, captivated by the twinkling expanse of the cosmos. This cosmic curiosity is more than just idle wonder; it's an innate drive for exploration, a yearning to push beyond the familiar and discover the unknown. However, this desire for discovery is now intertwined with a pressing need. As our population grows, Earth's resources become increasingly strained—a stark reminder that our planet is not an infinite wellspring. Space exploration, once the realm of our wildest dreams, is no longer a whimsical pursuit; it's a critical endeavor for securing humanity's long-term future.

The vastness of space holds the potential to alleviate pressure on Earth's resources. We have the opportunity to extract water, energy, and mineral resources from our immediate solar system. We not only have a pressing need to access extraterrestrial resources to supply our growing needs today, but we also have an obligation to future generations to do so sustainably and ethically, learning from the mistakes of past generations. We also have to consider expansion beyond Earth as a solution to overpopulation or crowding, as a way of improving our quality of life for our species in the future.

However, when it comes to exploring new worlds ourselves, the challenges of space travel are immense. The human body, a marvel of adaptation on Earth, becomes fragile in the harsh environment of space. Microgravity weakens bones and muscles while radiation bombards astronauts with the potential for long-term health problems. The psychological toll of confinement and isolation during long-duration missions adds another layer of complexity. These challenges necessitate not just technological advancements but a fundamental shift in our relationship with the cosmos.

Here, the philosophies and religions that have guided us on Earth may need to evolve. The vastness of space demands a new perspective on humanity's place in the universe. Acknowledging our place as a single species venturing into the vast unknown, perhaps we will forge new belief systems that embrace new views of religion and ethics. This shift in perspective is crucial for fostering a sense of shared purpose and collective responsibility, which are essential for the long-term success of space exploration endeavors.

AI, synthetic biology, and transhumanism are poised to play a transformative role in our journey to the stars. AI algorithms can analyze vast amounts of data to optimize spacecraft design, plan complex trajectories, and even assist astronauts with decision-making during critical missions. As AI capabilities continue to grow, we can expect them to become even more integral to all facets of space exploration. Synthetic biology allows us to engineer organisms with specific functionalities, like microbes designed to biomanufacture food, fuel, and even building materials on the surface of Mars or other celestial bodies. Transhumanism, on the other hand, envisions augmenting human capabilities to withstand the rigors of space travel. Genetic engineering holds the promise of enhancing human tolerance to radiation, microgravity, and other space-related stressors. These technologies, while still in their early stages, represent exciting avenues for redefining what it means to be a spacefarer.

Our world faces an array of daunting challenges, from climate change to social inequality, and we will need to expand to the cosmos to address these issues. This can seem overwhelming, but instead of succumbing to despair, we should embrace them as opportunities for growth and innovation. As Bobby Azarian argues in *The Romance of Reality*, problems are the engines of progress. "Yes, the world has some serious problems," Azarian writes, "but if we did not have problems, we would never be forced to find new solutions. Problems push progress forward."[1] The process of solving complex challenges forces us to transform raw information into valuable knowledge, a fundamental driver of evolution. By tackling these issues head-on, we're not just fixing problems; we're actively participating in the universe's relentless pursuit of greater complexity and intelligence.

The road to a future in which humanity thrives among the stars will be paved with innovation. Throughout this chapter, I'll discuss some of the problems that seem so insurmountable, we'll need moonshot moments to succeed. I'll do my best to highlight some of today's startups that are embracing that challenge and looking to make those moonshots now. The journey to the cosmos is not without its challenges, but the potential rewards are immeasurable.

Our journey begins with a critical question: Why should we venture into the vast unknown of space? The first half of this chapter will answer that call, exploring the compelling reasons why space exploration is not just a whimsical dream but a necessary step for humanity's future. The latter half will then shift focus, diving deep into the practicalities. I'll examine the tools we need—from philosophies to tech—and explore the pivotal moonshot moments that will be key to unlocking our cosmic destiny.

Ultimately, humanity's journey into the cosmos may mark an end to life on Earth as we know it, but I believe we are in fact at the beginning of a dream long ruminating in the backs of our minds, buried long ago by our ancestors.

Water and Energy: Overcoming Scarcity with a Win-Win Mindset

Our species thrives on a seemingly endless bounty. Fresh water sustains our crops, fuels our industries, and quenches our thirst. Energy, the lifeblood of progress, powers our homes, lights our cities, and propels us forward. Minerals, the building blocks of our modern world, are woven into everything from smartphones to skyscrapers. But like any grand feast, the resources that fuel our existence are not limitless. Here on Earth, the banquet tables are groaning, and cracks are starting to show.

Consider the silent crisis unfolding in our breadbaskets. Farmers in California, once the envy of the world, now grapple with water shortages that threaten their crops.[2] What were once dependable rivers are now a trickle, stark reminders of our dependence on a finite resource. This isn't an isolated incident. Around the globe, cities are facing water rationing,[3] a chilling glimpse into a future in which the lifeblood of agriculture becomes a luxury.

Moons like Europa, a celestial pearl orbiting Jupiter, harbor vast subsurface oceans,[4] estimated to hold more water than all of Earth's oceans combined.[5] These hidden reservoirs, encased in a layer of icy crust, are a tantalizing glimpse into the potential bounty locked away in our solar system. Asteroids, those seemingly barren space rocks, can also be surprisingly water rich. Certain types of asteroids, known as C-type asteroids, are believed to contain a significant percentage of water ice,[6] potentially a more accessible source than the subterranean oceans of Europa.

Several companies are vying for a leading position in this crucial endeavor. Argo Space Corporation, for instance, is developing technology to extract water from the moon's surface and redirect it to missions already in space, as fuel.[7] Meanwhile, Lunar Outpost is taking a more direct approach. Their vision involves building a new type of rover with water extraction as a core functionality. Lunar Outpost's Heavy In-situ Propellant Production Off-world Rover (HIPPO) utilizes AI and robotics to locate and extract water ice from craters on the Moon.[8] This extracted ice could then be processed into drinkable water and breathable oxygen for lunar inhabitants.

Masten Space Systems, renowned for their work on reusable rockets, is also entering the lunar water market. Several of their lunar landers are being designed with the capability to integrate water harvesting technologies.[9] This versatility could make the landers a valuable tool for future lunar expeditions, allowing them to land, harvest water ice, and potentially even refuel with extracted hydrogen. Finally, LIQUIFER Systems Group brings their expertise in microwave extraction to the table. Their Lunar Water Extraction System (LUWEX) utilizes microwaves to heat lunar regolith, vaporizing trapped water ice and other volatiles.[10] This innovative approach could pave the way for efficient and scalable water extraction on the Moon. With these companies at the forefront, the dream of finding and utilizing water in space is fast becoming a reality.

The implications of these technologies are staggering. Water, not just essential for life as we know it, but also a valuable resource for everything from drinking water to fuel production, could be harvested from these celestial bodies, alleviating our dependence on Earth's dwindling freshwater supplies and ensuring a sustainable future for generations to come. The vastness of space may seem daunting, but the potential for water ice on moons and asteroids offers a glimmer of hope, a reminder that the universe might just hold the key to unlocking a future in which water scarcity becomes a relic of the past.

Energy, the engine driving our civilization, presents a unique challenge. The fossil fuels that have powered our growth for centuries come with a heavy price tag—a warming planet and a polluted atmosphere. We turn to renewables—solar, wind, geothermal, and hydro—but these solutions, while promising, have limitations. Solar panels are becoming more efficient, but their large footprint limits their widespread adoption. Wind turbines are impressive feats of engineering, but their effectiveness depends on unpredictable wind patterns. These limitations highlight the need for a technological leap, a breakthrough that unlocks the full potential of renewable energy sources.

Here, on Earth, our access to our sun as a celestial power source is constantly interrupted by the pesky veil of our atmosphere. Clouds scatter and absorb sunlight while the rotation of the planet plunges us into the periodic darkness of night. But beyond the embrace of our atmosphere lies a potential game-changer: solar energy in its purest, most potent form.

Space-based solar power offers a revolutionary approach to energy generation. Freed from the limitations of Earth's atmosphere, solar panels in space can capture the sun's rays with exceptional efficiency because there are no clouds to cast shadows, no dust to dim the light—just a constant stream of clean, uninterrupted energy. The potential is truly staggering. Could we see vast solar arrays in medium Earth orbit, constantly basking in the sun's brilliance and beaming clean energy back to Earth wirelessly?[11] This is the ambitious goal of Space Solar, a UK-based company leading the charge. Space Solar's team argues that these arrays wouldn't just supplement our existing energy grids but could revolutionize them.[12] We could transition away from fossil fuels, the silent culprits of climate change, and embrace a future powered by the boundless energy of our nearest star.

Other companies are also involved in the space energy game. Orbital Composites is a space manufacturing startup that's playing a crucial role in building the infrastructure for space-based solar power.[13] They specialize in creating composite materials specifically designed for the harsh environment of space. These lightweight, high-performing materials are essential for building the massive structures needed to collect solar energy in space. With their innovative approach, Orbital Composites is helping to pave the way for a future powered by the sun, even from beyond Earth's atmosphere.

Virtus Solis, a space energy company that is laser focused on developing the satellite array technology to make space-based solar power a reality, aims to pilot its space-based solar tech in orbit in 2027.[14] Orbital Composites and Virtus Solis are also teaming up to construct a whole constellation of satellites that would beam the energy produced by Virtus Solis back to the ground.[15] This is the kind of hyper-collaboration that will be vital to making space solar power a reality.

The challenges of space-based solar energy, of course, are significant. The cost of launching and maintaining these solar arrays in space is currently high. Technological advancements are needed to develop lightweight, efficient solar panels and satellites that can withstand the harsh environment of space. But the potential rewards are immeasurable. A shift to space-based solar power wouldn't just provide us with a clean and sustainable energy source, it could also pave the way for a future in which energy independence becomes a reality for nations around the globe.

Right now, though, the harsh reality is simple: our current consumption patterns of water and energy on Earth are unsustainable. This does not mean that we should panic—remember, we are a species that thrives on win-win scenarios and many of our worst moments as a species have been when we've prioritized a scarcity mindset. While being aware—and respectful—of the scarcity of resources on Earth, we should not let that be the focus. We must focus on innovation and expansion rather than quibble over resources.

Sharing Mineral Wealth: Yours and Mine

The road to artificial general intelligence might seem paved with groundbreaking algorithms and elusive breakthroughs. However, the current bottleneck lies not in the realm of pure innovation but in the very nuts and bolts of computing power.[16] Training the complex neural networks that hold the promise of AGI requires a staggering amount of processing power. This, in turn, translates to a very real-world limitation: the sheer number of chips companies like Nvidia, Taiwan Semiconductor, and others can produce. The next great war between the United States and China may be over control of the chip production industry.[17]

These chips, the workhorses of AI development, are themselves marvels of engineering, demanding a complex dance of materials science and intricate manufacturing processes. Unfortunately, the resources needed to create these chips can be just as scarce as the processing power they deliver. In particular, the chips used in operating AI require rare earth minerals, primarily mined by China[18] and, increasingly, Australia.[19] Obtaining these rare earth elements and other materials necessary for chip construction can be a logistical hurdle, further slowing down the pace of progress.

The explosive growth of Nvidia, TSMC, and other manufacturers since late 2023 also shows that we will see power and wealth coalesce in the hands of the few companies that produce these chips or mine the resources for them. In essence, the key to unlocking AGI might not be a mind-blowing discovery but rather a scaling up of what we already know how to do—building more powerful and readily available hardware.

In many ways, the finite supplies of certain minerals and rare elements on Earth may be the strongest catalyst for pushing the exploration of space, at least within our solar system. Our first permanent settlement beyond Earth likely won't be a terraformed moon or Mars, but manned extraction stations on near-Earth asteroids between Mars and Jupiter. This isn't science fiction; it's the burgeoning field of space mining, and though we are still far away from returning investment on the space mining industry,[20] it is necessary for the funding community to support wild thinking and moonshot innovation. Startups like Asteroid Mining Corporation, AstroForge, and TransAstra are at the forefront of this endeavor, developing innovative technologies to extract these valuable resources from asteroids.

Asteroid Mining Corporation takes a theoretical approach, focusing on research and development of core technologies for asteroid exploration and resource extraction. They're working on satellite platforms that can scan and analyze near-Earth asteroids, identifying potential targets and building the knowledge base necessary for future mining operations.[21] They are essentially focusing on the first step of data acquisition, which can allow other companies to then focus on the technology needed to extract, refine, and transport mineral wealth from asteroids back to Earth.

AstroForge focuses on a critical aspect of space mining—in-space refinement.[22] Their approach tackles the logistical hurdle of transporting raw materials back to Earth for processing. Their technology utilizes solar energy to power a process that transforms asteroid ore into usable metals right there in space. This not only reduces the immense cost of transporting bulky materials but also allows for a more targeted selection of the most valuable elements. We may see a future in which spacecraft, constructed from space-mined materials, are built in orbit, ready to explore the cosmos with a lighter footprint. TransAstra, on the other hand, focuses on the transportation aspect of space mining. They're developing powerful and efficient space tugs that can be used to move mined resources or even entire mining spacecraft around the solar system.[23] Their technology could play a crucial role in making space mining a truly sustainable endeavor.

These are just a couple of examples of the innovative companies leading the charge in space mining. Their advancements hold the promise of not only mitigating resource scarcity on Earth but also fueling the next generation of space exploration and technological innovation. It's a new frontier, and the first prospectors might not be wielding pickaxes, but they'll be paving the way for a future built on the riches of the cosmos.

The potential of space mining is vast, but it's crucial to remember that the cosmos is a shared resource. While the companies mentioned here represent the cutting edge of this endeavor, international collaboration will be paramount. Sustainable space mining that benefits all of humanity, not just a single company or nation, requires open communication, shared knowledge, and a commitment to responsible resource utilization. The future of space exploration hinges on developing win-win solutions, ensuring that the riches of the cosmos fuel not just technological progress but also global cooperation and a brighter future for generations to come.

Home, Sweet Home

And so humanity, with its remarkable capacity for growth, has pushed Earth's resources to their limits. The concept of carrying capacity reminds us that any environment can only support a finite number of organisms. Fresh water, a vital resource for all life, is becoming increasingly scarce, and will limit population growth and expansion. Our agricultural lands are strained. Climate change, fueled by our dependence on fossil fuels, further complicates the picture. Rising sea levels threaten coastal communities, and extreme weather events disrupt agricultural production. These are not distant threats; they are realities we face today.

The relentless march of human population growth, currently estimated to be over 8 billion and projected to keep rising, intensifies the pressure on these already strained resources. AI-improved medical research and longevity studies will prolong our lifetimes as well, adding more population pressure. While some futuristic visions would solve this by transcending our physical forms, I again argue that embracing brain uploading seems far-fetched for most. Our bodies are a fundamental part of who we are, shaping our experiences and interactions with the world. This innate connection to our physical selves suggests that space exploration, not abandoning our bodies, is the more likely path for humanity's long-term survival and expansion. Space exploration offers the key to securing a future for our embodied existence, not just our minds, among the stars. Thus, we do not just need more resources from the cosmos—we also need room to physically expand.

While challenges abound on Earth, humanity has never gladly accepted its limitations. Space exploration offers a potential solution, which we should not view simply as an escape but instead view as an expansion. The vastness of space holds the promise of resources we can only dream of here on Earth. Beyond our solar system, exoplanet research has identified thousands of potential candidates, some within habitable zones where liquid water—a key ingredient for life as we know it—could exist. While we haven't definitively found another Earth yet, the sheer number of possibilities underscores the potential for finding a new home for humanity among the stars.

Indeed, as humanity sets its sights on the cosmos, our concept of "home" must evolve. No longer will it be confined to a single planet, a collection of countries, or even a cozy house. Our future home will encompass the vast expanse of space, a network of colonies and habitats scattered among the stars. This doesn't diminish the importance of our Earthly home; it broadens the definition, embracing the potential for new worlds to become havens for future generations, ensuring the continuation of the human story in the grand narrative of the universe.

Of course, venturing into the cosmos comes with its own set of challenges. Concerns about resource exploitation in space are valid. We must ensure sustainable practices are in place to avoid repeating the mistakes we've made on Earth. International treaties and regulations are crucial to govern space exploration responsibly, preventing any one nation from claiming entire celestial bodies or exploiting them for their own gain. The potential for environmental damage, such as space debris colliding with satellites, also needs to be addressed. These challenges are not insurmountable. The history of scientific advancement demonstrates our ability to develop solutions to complex problems.

Some might argue that the resources needed for space exploration could be better spent solving problems here on Earth.[24] This argument misses the bigger picture. Space exploration is not a luxury; it's an investment in our long-term future. The solutions and technologies developed for space exploration have the potential to address many of the challenges we face on Earth, from developing new energy sources to improving food production techniques. The knowledge gained from studying other planets will inform our efforts to combat climate change and ensure the sustainability of our own.

Additionally, investing in space exploration isn't just about finding a new place to live, it's about fostering innovation and economic growth right here on Earth:[25] even as the long-term goal seems far off, there is value in spending resources in the present to work toward future space exploration. The challenges of space travel have historically pushed the boundaries of human ingenuity. The development of lightweight and heat-resistant materials for spacecraft has led to advancements in fireproofing and construction technologies.[26] Medical research conducted for astronauts on long-duration missions has yielded breakthroughs in areas like bone health and wound healing.[27] The nascent space economy holds immense potential. Space tourism, while still in its infancy, could become a multibillion-dollar industry. International collaborations on space projects like the International Space Station (ISS) demonstrate the unifying power of a shared goal, fostering cooperation and diplomacy between nations. Private space ventures, like those spearheaded by Elon Musk and Jeff Bezos, are further accelerating the pace of innovation and exploration.

While space exploration offers a solution for our long-term sustainability, the reasons to venture beyond our home planet extend far beyond mere necessity. We are, by nature, a species driven by an insatiable curiosity, a yearning to push past the known and explore the unknown. The settlement of new worlds is the ultimate expression of our inherent human desire to explore, a pursuit that has defined our history and will continue to shape our future among the stars.

Our Prime Directive

Space exploration transcends resource acquisition or territorial expansion; it's a call to unravel the cosmic mysteries, to discover new worlds and wonders that ignite our imagination. It's this inherent desire that compelled our ancestors to cross vast oceans, climb the highest mountains, and dig into the deepest parts of our planet. Space exploration represents the logical extension of this primal urge. This pursuit of knowledge and understanding is a fundamental part of what makes us human, and it's a driving force behind our continued exploration of the universe.

It's easy to look at earlier periods of human history to see how we have handled exploration before. The "Age of Exploration," roughly spanning the fifteenth to the seventeenth centuries, was a period driven by a potent mix of ambition and necessity. The fall of Constantinople to the Ottomans in 1453 choked off traditional land routes to the riches of Asia, pushing explorers westward across the Atlantic and southward around Africa in a zero-sum game of politics and trade. European nations, particularly Portugal, Britain, France, and Spain, craved new trade routes to Asia, fueled by the allure of spices, silks, and other riches. Technological advancements like the compass, astrolabe, and more durable ships also emboldened these voyages. Entirely by accident, the "New World" (which was already well populated by technological and socially advanced peoples) was "discovered," changing the course of human events.

The Age of Exploration's outcomes were vast and complex. On the positive side, it led to the discovery of new continents, the exchange of plants, animals, and ideas among civilizations, and a surge in global trade, all of which would have modern equivalents as we reach out to space. However, the era was also marked by brutal colonialism, the exploitation of indigenous populations and local resources, and the spread of diseases that devastated native communities. We would be foolish in the extreme to ignore such consequences as we push forward into space.

Ultimately, however, the Age of Exploration was just a manifestation of a naturally selected penchant for exploration and expansion. *Homo sapiens*, our species name, translates to "wise man," but perhaps "wise explorer" would be equally fitting. Our evolutionary story begins with our ancestors leaving the familiar comfort of Africa's savannas, venturing into diverse and unpredictable environments. This wasn't a one-time event; it was a continuous push outward, driven by a relentless curiosity about what lay beyond the next horizon. Those who thrived weren't necessarily the strongest but the most adaptable and hyper-cooperative. It's entirely possible that the first complex communication did not derive from complex emotions, but from the need to share directions, resources, and strategies for survival in these new frontiers. Our very biology, shaped by generations of explorers, may be primed for the ultimate exploration—venturing out into the cosmos itself.

No cultural artifact, fictional or factual, encapsulates this spirit of human exploration and discovery more perfectly than Gene Roddenberry's *Star Trek* universe. It's mantra—"to explore strange new worlds; to seek out new life and new civilizations; to boldly go where no man has gone before!"—capture the very energy I'm describing. Of course, we've now moved beyond "where no man has gone before" to include people of all genders, but perhaps that will not even be sufficient as we bring AI along for the ride. But, slogans aside, the *Star Trek* universe has come to represent, now for several generations, the very best of humanity's spirit of discovery and pure wonderment. Not only that, but many of today's scientists, astronauts, and technologists—especially Black women, thanks to the truly groundbreaking Uhuru character in the 1960s original series[28, 29]—have directly attributed their careers in tech to watching the various *Star Trek* series and films.[30]

That's something I have always found interesting, not being a Trekkie myself: the *Star Trek* universe has inspired such a wide range of innovators I meet today in the tech field, especially those who watched *Star Trek: The Next Generation* and then grew up to become the titans of the dot.com boom. "TNG," as it's often called, aired from 1987–1994 and featured notable main characters like a blind Black Chief of Engineering and a fully sentient AI android struggling to experience emotions. Both characters were revolutionary in the ways they inspired the next generation of scientists and research. Perhaps the most interesting crew member, however, is the captain himself—not a military strategist or engineer, but an anthropologist and diplomat. Whether deciphering an alien language based entirely on mythology, or arguing in court that the android was indeed an entity with the same full rights as a biological being, Captain Jean Luc Picard taught an entire generation of tech minds that our push into the cosmos was a cultural and moral journey, and an opportunity to be more than just a resource collector and alien conqueror.

While the vastness of space offers the potential for resources and room for humanity to grow, the journey to get there requires more than just spaceships. We need not only the technological tools for exploration, which we'll explore next, but also a new set of guiding principles. The Prime Directive from the original *Star Trek* series was a philosophy that banned interference with any civilization not yet capable of faster-than-light space travel, which demonstrates the need for us to develop guiding principles and philosophies that will accompany humanity on its journey to the cosmos.

New Philosophies: Technomoral Virtues in Space

As the ethical considerations of space exploration come to the forefront, the concept of technomoral virtues that we've already discussed[31] offers a valuable framework for navigating the complex challenges of spacefaring. We need to develop these virtues first for working with AI, synthetic biology, robotics, quantum computing, and other advanced technologies here on Earth. But there will need to be more, and we cannot wait until we're in the cold of space to start considering the philosophical systems we want to apply throughout the cosmos. In the vast expanse of space, the application of humanity's technomoral virtues takes on a whole new dimension. Extending these technomoral virtues to space exploration is crucial for ensuring responsible and ethical practices as we venture beyond our home planet.

There are some specific fields where developing a robust ethical framework for space activities will be critical. Resource extraction, a potential cornerstone of space expansion, necessitates clear guidelines to ensure its sustainability. We must avoid depleting resources on celestial bodies in a way that jeopardizes future generations' ability to benefit from them. Space debris, a growing problem in Earth's orbit, must also be addressed. Collisions with debris can cripple satellites and endanger future space missions. International collaboration and treaties are essential to establish responsible practices for waste disposal and debris mitigation in space.

The concept of terraforming, modifying a planet or moon to make it resemble Earth and support human life, raises profound ethical questions. While the prospect of creating a new home for humanity on another world is undeniably captivating, we must tread carefully. Does terraforming constitute an act of arrogance, imposing our will on another celestial body? Do these planets or moons harbor existing ecosystems, even if they are microbial? Respecting the potential for life beyond Earth and preserving the natural state of celestial bodies needs to be a central tenet of our spacefaring ethics.

Perhaps the most mind-boggling ethical considerations surround the possibility of encountering alien life. Personally, I am a believer in the Fermi paradox—the fact that we haven't heard from intelligent life yet likely means there either is none or it's too far away to interact with.[32] As Max Tegmark argues, "I think that this assumption that we're not alone in our universe is not only dangerous but probably false."[33] Nick Bostrom, a philosopher concerned with existential risks, uses the concept of a "Great Filter" to address the Fermi paradox.[34] Bostrom suggests that the development of complex life might be a rare event, with numerous hurdles or bottlenecks acting as a kind of "Great

Filter." These bottlenecks could be physical limitations like the emergence of complex life from prebiotic soup, or late-stage developments that block galactic travel. If we had already met a species transversing the stars, we might assume bottlenecks come early in our development and we've already successfully made it through, but we haven't met them yet.

So, Bostrom posits, if we find low-level life on other planets, we must come to terms with the fact that the Great Filter comes late in development, meaning that upcoming challenges in technology or social development are the factors limiting growth. By proposing the Great Filter, Bostrom offers a possible explanation for the lack of alien civilizations despite the vastness of the universe. He encourages us to think that "we would have some grounds for hope that all or most of the Great Filter is in our past if Mars is found to be barren. In that case, we may have a significant chance of one day growing into something greater than we are now."[35] Our own existence, then, becomes a testament to overcoming the Great Filter, but it doesn't guarantee the prevalence of intelligent life elsewhere. For these reasons alone—that we might very well be alone, facing upcoming species-wide challenges—humanity must do everything it can to preserve itself and expand as far as possible into the universe.

Bobby Azarian paints an alternative, but no less sweeping, vision of humanity's cosmic destiny. He posits that we are not mere accidents of evolution but integral players in a grand cosmic drama. Our ability to process information and build complex internal models is a testament to the universe's relentless drive toward complexity and consciousness. Gaining more knowledge and spreading out into the cosmos to expand our knowledge is, he argues, a moral imperative.

> We, as sentient beings, truly are engaged in a great cosmic battle, one
> that quite amusingly turns out to be a story not unlike the eternal
> wars between good and evil as described by the world's major re-
> ligions. But the cosmic "struggle for existence," the challenge that
> faces any sentient species anywhere in the cosmos, is more like the
> one Deutsch described—a grand and majestic war between order
> and chaos, life and entropy, existence and nonexistence. Whether

one is religious, atheistic, or agnostic, we are all by default engaged in this spiritual war. I mean this in the most literal way; sentience can only continue to exist—feeling and experience and meaning and purpose can only continue to be part of physical reality—if life can persist in the face of the second law [of thermodynamics], and the only way it can do that is by acquiring knowledge.[36]

Azarian suggests a future in which humanity could evolve into something far beyond our current understanding, perhaps even merging with technology to create a planetary or even cosmic mind. He even posits the idea that our intelligent choices during the growth process could split off multiple multiverses,[37] and it is our free will and choices that shape the form of our universe. However, Azarian cautions that this future is not guaranteed. It requires conscious effort, global cooperation, and a deep understanding of our role in the cosmic scheme. Most importantly, it requires us to always be seeking new knowledge and learning: a new version of the Fermi paradox imperative that might suggest that knowledge is our way past Bostrom's Grand Filter.

On a longer-term scale, we have to consider the possibility that humanity could encounter other life-forms. And if we discover intelligent extraterrestrial beings, a whole new set of questions arise—ones we should not wait to answer until someone's knocking at the door. How do we as a species communicate with alternative life-forms? What protocols should guide our interactions? What knowledge do we want to share with others? Should we prioritize first contact, or is it ethically sound to observe them from afar, allowing their civilization to develop without our influence? Developing ethical protocols for communication and interaction with extraterrestrial intelligence will be paramount, demanding a level of empathy and understanding that extends beyond our current frame of reference.

As artificial intelligence takes over the mundane tasks that once filled our days, a unique opportunity emerges. The very automation that frees us from the shackles of repetitive labor allows us to turn our minds to loftier pursuits. One such pursuit is the crucial development of a robust ethical framework for space exploration. The vastness of space beckons, but venturing out requires a moral compass, a set of technomoral virtues that will guide our interactions with the unknown. The time freed up by AI can be a springboard for philosophical exploration, ensuring that, as we explore the cosmos, we do so with a deep sense of responsibility and respect for the wonders that await us.

New Religions and Mythologies

As humanity ventures beyond Earth and into the vast unknown of space, a profound question arises: Will this exploration give rise to entirely new belief systems and mythologies? As I mentioned earlier, advances in artificial intelligence and synthetic biology raise profound questions about the future of work and the very nature of death. These advancements may necessitate the development of new philosophical frameworks and rituals to provide meaning and purpose in a world potentially reshaped by automation and extended lifespans. Space exploration, however, offers a different path. Instead of grappling with the existential implications of a technological future on Earth, space beckons us outward, inviting us to confront the mysteries of the cosmos and rekindle a sense of awe and wonder in the unknown. Exploration should be a chance to create new narratives and belief systems inspired by the vastness of space, not a response to the challenges posed by technological advancements on a finite planet.

Throughout history, our understanding of the universe and our place within it has profoundly shaped our religions and myths. From the ancient Egyptians worshipping the sun god Ra to indigenous cultures weaving stories about constellations, the cosmos has served as a constant source of wonder and inspiration for spiritual beliefs. As we explore space and encounter the unfamiliar, it's reasonable to expect a similar creative response, potentially leading to the birth of new religions and mythologies.

Just as the vastness of the ocean or the mysteries of the night sky ignited the imaginations of our ancestors, space exploration offers a new frontier for sparking human creativity. What will our imaginations come up with when humans first witness a breathtaking gas giant with swirling storms unlike anything seen on Earth, or discover a celestial body with a double sunset? These awe-inspiring experiences have the potential to evoke a sense of wonder and a yearning to understand the forces that govern the universe. This could manifest in the creation of new narratives, similar to how our ancestors used myths to explain natural phenomena they couldn't comprehend. Perhaps spacefaring cultures will develop stories about the first human to walk on another planet, or weave tales about the origins of life in the cosmos.

The possibility of encountering life beyond Earth, whether intelligent or microbial, could have a particularly significant impact on our belief systems. Are we truly alone in the universe, or are there other intelligent beings out there? Do they possess their own belief systems, or is the concept of a higher power universal? This could lead to a reevaluation of existing religious doctrines or even a complete paradigm shift in how we perceive the divine. Space exploration has the potential to broaden our perspective, fostering a more interconnected and universal view of existence. We may move away from an Earth-centric understanding of the cosmos and embrace a more inclusive narrative that acknowledges the vastness and diversity of the universe.

The challenges and triumphs of space exploration itself could become the foundation for new myths and legends. The bravery of the first astronauts venturing beyond Earth's atmosphere, their courage immortalized in stories passed down through generations, would become the next Odysseuses and Gilgameshes. The struggles and sacrifices made in establishing a permanent human presence on another planet could be woven into narratives that inspire future generations of spacefarers. These stories would not only celebrate human achievement but also serve as a reminder of the unwavering human spirit and our inherent desire to explore the unknown. Space pioneers and astronauts could become revered figures in new belief systems, their accomplishments taking on an almost mythical quality.

But bold leaps are never without pain. Venturing into the uncharted territory of space-based religions also presents potential challenges. Existing religions with established doctrines and belief systems may view these new narratives with skepticism or even hostility. It's important to acknowledge these concerns and foster a spirit of tolerance and open-mindedness as we explore the cosmos. Perhaps space exploration can lead to a more peaceful coexistence between different belief systems, encouraging a shared sense of wonder and a collective pursuit of knowledge about the universe. Perhaps space exploration can be the win-win that connects all the philosophies of different faiths.

Space exploration has the potential to significantly impact the landscape of human belief systems. New religions and mythologies may emerge, inspired by the awe-inspiring beauty and mysteries of the cosmos. The possibility of encountering life beyond Earth could fundamentally challenge our understanding of divinity, and the challenges and triumphs of space exploration itself could be woven into new myths and legends. While potential conflicts with existing religions need to be addressed, space exploration also presents an opportunity for fostering tolerance and a shared sense of wonder about the universe. Ultimately, venturing into the cosmos may lead to a diversification of human belief systems, fostering a more interconnected and wonder-filled view of the universe, where humanity's place is just one chapter in the grand cosmic story.

Finally, we've explored the compelling reasons for humanity to set sail among the stars, contemplated the philosophical and religious frameworks that will guide us on this cosmic voyage, and acknowledged the ethical considerations that demand our attention before we undertake this next age of exploration. But before we boldly venture into the unknown, a crucial question remains—what tools will propel us beyond our home planet? While the intricacies of rocketry and the engineering marvels needed for long-distance travel are fascinating topics, for those details, we can turn to the visionary minds of Musk's and Bezos's teams, as well as the groundbreaking work being done in rocket programs at NASA, in Russia, China, India, and other countries.

Instead, let's take a peek into a different kind of engine room, one fueled by the potent intersection of artificial intelligence, synthetic biology, and the intriguing prospect of transhumanism. These advancements hold the potential to revolutionize space travel, paving the way for a future when humanity's reach extends far beyond the cradle of Earth.

Artificial Intelligence: The Mind Behind the Mission

Artificial intelligence in space exploration—the phrase might conjure up images of HAL 9000, the murderous computer from *2001: A Space Odyssey* (1968), sparking anxieties about intelligent machines gone rogue. Science fiction films like *Alien* (1979), *Moon* (2009), and *Passengers* (2016) have further cemented this fear in popular culture. However, these anxieties belie the reality—AI is already a vital, and well-behaved, partner in space exploration. AI's capabilities extend far beyond automation; it has the potential to revolutionize space exploration by assisting with navigation, decision-making, and scientific discovery. Let's dive into the ways AI is currently transforming our understanding of the cosmos and how it will continue to propel us further into the unknown.

When I close my eyes, I can easily imagine a spacecraft venturing deep into the vast expanse of space millions of kilometers away from Earth. But I also imagine that the challenges of navigating such distances with limited human intervention are immense! Here, AI-powered autonomous navigation systems come to the forefront. These systems, developed by companies like Lockheed Martin[38] and Airbus,[39] can meticulously plan and execute efficient routes, taking into account factors like gravitational forces, celestial object positions, and fuel consumption. Designed for Earth-orbital aircraft, these systems will likely form the basis of advanced space AI navigation systems. Furthermore, AI will be used to detect and avoid celestial hazards like asteroid fields or radiation belts, ensuring the safety of the spacecraft and its crew. On the surface of celestial bodies, AI-powered rovers and probes can play a vital exploration role in-situ.

The vast deluge of data from space missions presents another challenge—extracting meaningful insights from this information overload. This is where AI excels. AI algorithms can analyze immense datasets from telescopes, space probes, and rovers, identifying patterns and anomalies that might escape even the most meticulous human scrutiny. For instance, the ExoMiner algorithm was specifically designed to analyze leftover data from the Kepler and Transiting Exoplanet Survey Satellite (TESS) missions, which were searching for faint signals that might indicate the presence of exoplanets, potentially habitable worlds, beyond our solar system.[40] Similarly, AI can be used to analyze data from probes orbiting celestial bodies, potentially identifying resources like water ice hidden beneath the surface. This information is invaluable for establishing sustainable outposts on other planets or moons.

Beyond scientific discovery, AI can also play a crucial role in resource management on long-duration space missions. Let's say that, in the future, humanity decides to build a self-sustaining habitat on Mars, where resources like fuel, water, and oxygen are limited. Here, AI can optimize resource allocation and consumption, ensuring the well-being of the crew within the constraints of a closed-loop system. Companies like Axiom Space are developing AI-powered systems for spacecraft life-support systems, monitoring and optimizing environmental controls and resource usage.[41] This not only ensures the safety and comfort of the crew but also paves the way for longer and more sustainable space missions.

It's important to remember that AI is not a replacement for human ingenuity; rather, it's a powerful collaborator. Human oversight and ethical considerations remain paramount when developing and deploying AI for space exploration. Decisions with far-reaching consequences should always involve human judgment, with AI providing critical analysis and support. However, AI can free astronauts from mundane and repetitive tasks, allowing them to focus on more strategic endeavors, like conducting scientific experiments, piloting spacecraft during critical maneuvers, or making first contact with extraterrestrial life.

The fingerprints of AI are present in many ongoing space missions. NASA's Mars rovers, Curiosity and Perseverance, utilize AI-powered obstacle avoidance systems to navigate the rugged Martian terrain.[42] These systems process visual data from the rovers' cameras in real-time, identifying and avoiding obstacles that could potentially damage the rovers. Similarly, the James Webb Space Telescope, a marvel of modern astronomy, utilizes AI to process the immense volume of data it collects,[43] helping scientists distinguish faint signals from distant galaxies and unlock the secrets of the universe.

Private space companies are also embracing AI. SpaceX is developing AI systems for autonomous landing of their Starship vehicles, potentially revolutionizing space travel by enabling reusable launch and landing capabilities. Algorithms derived during the Machine Learning 4 Search for Extra Terrestrial Intelligence initiative (ML4SETI) are being used to analyze data from radio telescopes like the Allen Telescope Array (ATA) in the search for other intelligent life in the cosmos.[44] These are just a few examples of how AI is transforming space exploration today, paving the way for a future when humanity's reach extends far beyond the stars.

Made of Stars: Transhumanism and the Evolution of Spacefarers

The human body is a marvel of adaptation, capable of thriving in a vast array of environments. However, the harsh realities of space travel push our biological limits to the breaking point. Microgravity wreaks havoc on bone density and muscle mass, leaving astronauts susceptible to fractures and weakened physical capabilities. Exposure to high levels of radiation during deep-space missions significantly increases the long-term risk of cancer and other health problems. The psychological toll of confinement and isolation cannot be ignored either, with astronauts experiencing anxiety, depression, and sleep disturbances during extended missions to the International Space Station. While countermeasures like exercise routines and radiation shielding offer some protection, the human body remains fundamentally unsuited for the long-term rigors of space travel. This is where emerging technologies like synthetic biology, implants, and other forms of transhumanism offer a glimmer of hope.

Microgravity, the absence of the constant downward pull we experience on Earth, disrupts the delicate balance that maintains our skeletal and muscular systems. A 2020 study documented a bone mineral density loss of 1.0–1.5 percent per month in the spines of astronauts spending six months on the International Space Station (ISS).[45] This translates to a significant risk of fractures, a potentially life-threatening condition in the confined environment of a spacecraft. Muscle atrophy poses another challenge. Astronauts on the ISS lose up to 20 percent of their muscle mass in the legs within their first two weeks in space, rising to 30 percent after three months.[46] Regaining this lost strength requires extensive physical therapy upon return to Earth, a process that can take weeks or even months.

The radiation environment in space is another major concern. Without Earth's protective atmosphere, space is an obstacle course of harmful radiation from solar flares and cosmic rays. Studies have shown that astronauts on a mission to Mars would be exposed to a lifetime dose of radiation exceeding the safety limits set by space agencies.[47] This radiation exposure increases the risk of developing various cancers, including leukemia and gastrointestinal cancers, as well as cataracts and other health problems. While radiation shielding offers some protection on spacecraft, it adds significant weight and volume, limiting mission capabilities.

The psychological impact of space travel cannot be ignored. Astronauts on the ISS experience a unique combination of confinement, isolation, and sensory deprivation. A significant amount of research has shown that astronauts on long-duration missions exhibit symptoms of anxiety, depression, and sleep disturbances.[48] The psychological strain becomes even more critical when considering the mental well-being of crews embarking on multiyear voyages to distant planets. Current countermeasures like exercise routines and limited social interaction through video calls offer some mitigation, but they are not enough. Addressing these challenges requires a more comprehensive approach, one that seeks to not just mitigate the negative effects of space travel but to potentially enhance human resilience for a future of deeper exploration.

The human limitations exposed by space travel necessitate a paradigm shift, and synthetic biology emerges as a powerful tool for forging a path forward. This field harnesses the power of genetic engineering to create novel organisms with specific functionalities. By applying these principles, we can envision a future in which custom designed microbes become invaluable partners in space exploration, mitigating logistical challenges and paving the way for a sustainable human presence beyond Earth.

One of the most pressing concerns in space exploration is self-sufficiency. Astronauts currently rely on supplies transported from Earth, a costly and resource-intensive endeavor. Synthetic biology offers a solution through the development of microbes capable of biomanufacturing essential resources in-situ, on extraterrestrial bodies. Companies like LanzaTech are already pioneering the use of engineered microbes to convert industrial waste gases into fuels.[49] Building upon this foundation, we can envision the creation of microbes specifically designed to utilize local resources on planets or moons, like carbon dioxide and water ice, to produce food, biofuels, and even building materials. This would dramatically reduce dependence on Earth-based supplies and enable longer-duration missions farther into the cosmos.

Nature itself offers a wealth of inspiration for engineering organisms suited for space. Extremophiles—microorganisms that thrive in extreme environments like hydrothermal vents or hyper-saline lakes—possess unique adaptations that could be harnessed for space exploration.[50] For instance, *Deinococcus radiodurans*, a bacterium renowned for its exceptional radiation resistance, could serve as a template for engineering microbes capable of withstanding the harsh radiation environment of space.[51, 52] Similarly, microbes adapted to microgravity environments on Earth, like certain single-celled organisms found floating in high-altitude clouds, could provide valuable insights into developing organisms that function efficiently in the absence of gravity. By studying and understanding the remarkable adaptations of extremophiles, we can unlock the potential to create new life-forms specifically tailored for the challenges of space travel.

Beyond basic resource production, synthetic biology has the potential to develop microbes for environmental remediation and resource extraction. Space agencies are already exploring the possibility of sending robotic missions to prepare Mars for human habitation, which would involve terraforming the Martian atmosphere to make it more breathable—a feat NASA says is not possible without significant technological innovation.[53, 54] Engineered microbes that can fix nitrogen from the Martian atmosphere or degrade toxic perchlorates found in Martian soil could play a crucial role in this endeavor, if we can find a moonshot solution. Additionally, microbes designed to leach out valuable resources like metals and rare earth elements from asteroid material could revolutionize space-based mining operations, leading to the development of new materials and technologies.

The potential of synthetic biology for space exploration is undeniable. However, it's imperative to address the ethical considerations associated with introducing engineered organisms into alien ecosystems. The inadvertent introduction of invasive species could have devastating consequences for indigenous life-forms, if they exist, on other planets or moons. Strict planetary protection protocols must be established and rigorously adhered to in order to minimize the risk of contamination. Furthermore, open scientific discourse and international collaboration are crucial to ensure the responsible development and deployment of synthetic biology tools in space exploration. By harnessing the power of synthetic biology, we can engineer a new generation of microbes that will not only support human space exploration but may potentially unlock a future in which humanity thrives among the stars.

I've already discussed the importance of transhumanism in elevating humanity to the point of moonshot moments, but we will also need transhumanism to help push us further. As we venture deeper into the cosmos, the harsh realities of the space environment necessitate a moonshot cluster of innovations to address how we approach human exploration. Transhumanism offers a compelling vision—a future in which human augmentation allows us to not only survive but thrive in the face of space travel's many challenges.

Genetic engineering stands as a powerful tool within the transhumanist arsenal. By manipulating our genetic code, we can potentially enhance human tolerance to the detrimental effects of space travel. Microgravity, for instance, leads to significant bone loss. Researchers have identified a gene (LRP5) that plays a crucial role in bone density.[55] By understanding the mechanisms of this gene, scientists could potentially develop gene therapies to strengthen astronauts' bones and mitigate the risk of fractures during long-duration missions. Similarly, genes associated with muscle growth and repair could be targeted to counteract the muscle atrophy experienced in microgravity.

Space radiation also poses a significant threat to astronaut health. Transhumanism offers the possibility of engineering DNA repair mechanisms to enhance cellular resilience against radiation damage. Research published in *Nature* explored the potential of CRISPR gene-editing technology to introduce mutations that enhance a cell's ability to repair radiation-induced DNA damage. While such applications remain in their early stages, they represent promising avenues for exploring ways to mitigate the long-term health risks associated with deep-space travel and radiation.

Beyond physical enhancements, transhumanism envisions cognitive augmentation through the use of neural implants. Astronauts equipped with brain-computer interfaces (BCIs) could directly link their neural activity to spacecraft systems. This could revolutionize space travel by enabling more intuitive control of spacecraft and robotic systems. For instance, a BCI could translate an astronaut's thoughts into specific commands for the spacecraft, streamlining complex maneuvers and reducing reliance on manual controls. Companies like Neuralink, led by Elon Musk, are actively developing BCIs with potential applications in various fields, and it would seem obvious for SpaceX to experiment with astronaut BCIs when the technology is fully developed.

The integration of technology with the human body raises serious ethical concerns. While discussing biohacking, I've talked about my concerns with the potential creation of a "designer" species, where genetic enhancements become a tool for social and economic stratification. If we do not pursue a democratic transhumanist path, unequal access to these technologies could exacerbate existing social inequalities, leading to a future in which space travel, humanity's only means of survival as a species after ruining Earth, becomes the exclusive domain of the privileged few.

Childbirth in space or while on space missions is a concern for which humanity has not even begun to gather data, but considering how dangerous childbirth is on Earth, I cannot imagine we are remotely ready to begin exploring the issue in space ethically. In the harsh environment of space, traditional pregnancy poses significant health risks for women. Radiation exposure, microgravity-induced bone loss, and the physiological challenges of childbirth are just some of those concerns. Yet, if we are to manage long journeys without faster-than-light travel, we will have to consider pregnancy and childbirth outside Earth's protective womb.

Ectogenesis, the artificial development of an embryo outside the female body, offers a potential solution by decoupling human reproduction from the biological limitations of the human body. This novel concept presents far-reaching implications for space exploration.

One of the most compelling benefits of ectogenesis for space colonization lies in mitigating the health risks associated with pregnancy in space. By developing embryos in a controlled environment, scientists could shield them from harmful radiation and ensure optimal growth conditions. This would not only safeguard the health of the women participating in space colonization efforts but also offer greater flexibility in mission planning by removing the biological constraints of human gestation on the available labor force.

Beyond mitigating health risks, ectogenesis has the potential to enable the establishment of multigenerational populations on distant planets. The ability to develop embryos ex-utero would allow for a more rapid population growth on newly colonized worlds. This fosters a sense of permanence and community, crucial for establishing a foothold beyond Earth. I believe we will progress to a future in which human settlements on Mars or other celestial bodies are not merely temporary outposts but thriving societies with generations born and raised among the stars.

However, the potential of ectogenesis is not without its complexities. Ethical considerations surrounding parental rights, informed consent, and the potential commodification of human reproduction require careful consideration. A serious obstacle to this technology will be the developmental process, as Evie Kendal notes that "experimental ectogenesis will be quite risky, and it is the resultant offspring who will bear the consequences of any errors in the process."[56] She goes on further to ask about the ethics of forcing genetic mutations on children in artificial wombs, without their consent, that lock them into specific roles, much like Huxley's *Brave New World*. "In the case of off-world settlers, some modifications might even make it impossible for offspring to return to Earth."[57]

As interstellar travel becomes a possibility, the limitations of the human body become stark. For journeys lasting millennia or in spacecraft of limited size, the idea of uploading human minds into a non-biological substrate—essentially creating digital copies—emerges as a radical but potentially necessary solution. Combined with ectogenesis, uploading minds to stable synthetic platforms or media could rapidly populate entire terraformed worlds. While I do not think most humans will embrace mind uploading, it could be viewed as a necessary tool that would allow for the transportation of human consciousness across vast distances for extended periods, pushing the boundaries of space exploration beyond the constraints of our physical form.

The transhumanist vision for space exploration does not envision replacing humanity with machines. Instead, it seeks to create a synergistic relationship with technology that empowers humans to push the boundaries of human exploration. By harnessing the power of transhumanism, we can pave the way for a future in which humanity—not merely as biological beings, but as augmented explorers—can venture farther and deeper into the cosmos, forever expanding the frontiers of human experience.

The future of human exploration of space presents a daunting challenge that gleams with the promise of transformative technologies. Synthetic biology offers the potential to engineer organisms that can not only sustain us but also transform the Martian landscape or mine asteroids. Transhumanism, on the other hand, envisions augmenting human capabilities to withstand the rigors of space travel and unlocking new avenues for human-machine collaboration through synthetic biology, brain-computer interfaces, ectogenesis, and more. However, embracing these advancements necessitates a measured approach and, again, the establishment of strong technomoral virtues and values. Open dialogues are crucial to ensure responsible development and deployment, addressing ethical concerns surrounding bioengineering and human augmentation as a part of space exploration. As we navigate these complexities, the potential for a future in which humanity thrives among the stars remains a beacon to our enduring curiosity and the boundless potential of scientific exploration.

A Win-Win View of Space

In the grand tapestry of space exploration, hyper-cooperation and international collaboration will be the golden threads that tie countless moonshot moments together. The vastness of the cosmos demands a collective effort, a shared vision that transcends national borders and even necessitates a multigenerational effort. By pooling resources and expertise, we can unlock the potential of space resources for the benefit of all humanity. I believe in a future in which the wealth of the cosmos empowers not just a select few but fuels innovation and sustainable development on a global scale. This collaborative approach fosters a sense of shared responsibility, where the resources of space become a common inheritance to be safeguarded and utilized for the betterment of humankind.

However, the path to space resource utilization is not without its challenges. Ethical considerations regarding environmental impact on celestial bodies and the potential for resource exploitation require careful deliberation and the development of agreed-upon technomoral virtues. Robust international frameworks and open scientific discourse are crucial to ensure responsible space exploration and resource management. We must tread lightly on these new frontiers, ensuring our actions benefit future generations and preserve the pristine beauty of the cosmos. Yet, tread we must: exploration is the mark of our species, and moonshots are what we were born to achieve.

The urgency of space exploration cannot be overstated. As we face the challenges of climate change and resource depletion on Earth, space offers a solitary beacon of hope. By securing access to extraterrestrial resources, we can ensure a sustainable future for humanity, safeguard our collective existence, and pave the way for a future in which humanity thrives among the stars. With the help of emerging technologies and potentially powerful partnerships with AIs, we can shoot for far beyond the moon. The time for action is now. Let us embark on this shared endeavor, driven by the spirit of exploration, with a win-win mindset and a commitment to the betterment of all humankind. The journey ahead will be paved with its own moonshot moments—ambitious endeavors that ultimately push the boundaries of space exploration and resource acquisition. The cosmos, a treasure trove of potential, awaits and it is our collective responsibility to un-lock its secrets for the long-term benefit of all life on Earth.

Sleepless Nights

I've decided, after taking you on this long but powerful journey through the choices and opportunities that will push humanity to the cosmos, that I should end our discussion by telling you what keeps me up at night.

It's not the fear of AI. Unlike some who view artificial intelligence with fear and suspicion, I find myself energized by its rapid development. While the concerns of techno pessimists and Luddites are not to be dismissed entirely, I choose to see AI not as a looming threat but as a powerful tool with the potential to liberate us from the shackles of the mundane. The true power of AI lies not in replacing us but in empowering us to reach our full potential. I feel the same way about synthetic biology, robotics, and advancements in transhumanist tech-nology; we are on a path upward and onward, toward progress and expansion to the cosmos.

Nor does the panic over scarcity leave me sleepless. I reject the pervasive scarcity mindset that casts a shadow over our future. We exist on a planet brim-ming with resources and, with innovation as our compass, we can overcome any potential strains. The vastness of the cosmos beckons, offering a future in which humanity can expand its reach and access to materials. Even here on Earth, the spirit of innovation is alive and well. Countless startups are working tirelessly to develop solutions for resource management and sustainability. By embracing in-genuity and collaboration, we can ensure a thriving human population, not just here on Earth but potentially across the cosmos. The future is not one of scarcity but of abundance, driven by our collective spirit of exploration and invention.

What keeps me up at night is the Fermi paradox: If the universe is teeming with life, wouldn't we be interacting with it by now? The lack of a significant contact suggests that the challenges of reaching such advanced stages are so immense, or self-destruction so common, that either no other intelligent life exists or, at the very least, civilizations sputter out before achieving interstellar communication. That prospect alone—that we are pushing ourselves to a limit it seems no other species in the universe has overcome—is incredibly daunting. Our loneliness comes at a great cost: we must solve our challenges on our own, and we have a powerful obligation to survive and thrive.

Daunting too is the responsibility that this idea represents. I am driven by the knowledge that we might be the universe's very best shot at intellectual transcendence and interstellar exploration. I think the universe put all its proverbial eggs in one basket—and we're that basket. That responsibility, knowing that we, as humanity, have to succeed for the sake of the universe is exciting. We are its most intelligent and capable steward, not of planet Earth, but of the knowable universe itself.

Chapter Ten
IT'S ALL UP TO US

This book began as a series of gnawing questions in the back of my mind: What will give us purpose once AI removes our most mundane tasks? How can we push the limits of ourselves as we build new selves in the digital realm? What if we embraced opening our minds? What does the Fermi paradox mean for us as a species?

What began as a modest exploration morphed into a sprawling intellectual odyssey. As I dug deeper, casting a wider net across various disciplines, the project's scope expanded exponentially. My PKM vault, initially a tidy repository of foundational texts, swelled into my own personal LLM. It became a living document, enriched with a diverse array of materials: cutting-edge research on transhumanism, thought-provoking documentaries delving into the world of psychedelics, and insightful interviews with leading academics from fields as disparate as neuroscience, philosophy, and mathematics. The more I discovered, the more questions arose, propelling me further into the uncharted territory of questioning humanity's potential.

The theme of moonshot moments appeared again and again, whether it was the pressing need for them to solve emerging problems, or the unprecedented potential for growth and progress we could have as a species if we could harness them for good. Throughout this book, I've tried to tie together the threads I've seen as most important, in the hopes that you'll be as excited about our species' potential as I am.

And on this journey, we've followed humanity through some of the greatest moonshot moments of our species and begun to question what comes next in our voyage beyond our current reality. We've been able to draw some powerful conclusions about moonshot moments.

- **Collaboration and diverse thought are the cornerstones of innovation.** Effective knowledge sharing and hyper-cooperation are essential for maximizing the impact of our collective efforts.
- **A culture that embraces failure and experimentation is crucial for fostering a fertile ground for innovation.** Rewarding risk-taking behavior encourages individuals and organizations to pursue ambitious goals without fear of repercussions.
- **Challenging the status quo is often a necessary precursor to transformative change.** By questioning established norms and paradigms, individuals and organizations can uncover new opportunities and create disruptive innovations.
- **Associational thinking and mind expansion techniques can unlock hidden potential for innovation.** By connecting seemingly unrelated concepts and exploring alternative perspectives, individuals can generate novel ideas and solutions to complex problems.
- **Moonshot moments are imperative for addressing humanity's most pressing challenges and propelling us into a future characterized by unprecedented progress and exploration.** These ambitious endeavors have the power to reshape industries, improve lives, and expand the boundaries of human knowledge.

Recommendations for Those on the Front Lines of Wild Thinking

After drawing these conclusions about moonshot moments, I had to pause. Was the journey done? I wondered what to do with all of these observations about moonshot moments, and how to turn them into a set of solutions to the pressing problems of our day. In earlier drafts, I followed the path of Suleyman, Tegmark, Azarian, Bostrom, and others, laying out a sprawling list of pragmatic recommendations oriented toward tech developers, policymakers, and political leaders. And most certainly, many of those ideas still have strong merit.

For example, I'd argue that humanity must urgently cultivate a global culture of hyper-cooperation, fostering strong academic-industry partnerships and implementing proactive regulations for ethical development and responsible use. Achieving safe and ethical technological advancement necessitates a multifaceted approach, including robust regulation, industry-academia collaboration, independent oversight, open dialogue, and active consumer engagement to ensure technology benefits all of humanity.

I believe that we need a revitalized civil society focused on creativity, critical thinking, and collaborative problem-solving while simultaneously developing new philosophies and moral leadership to guide humanity through this era of rapid change. This includes cultivating a surge of youthful optimism and engagement by revamping education, leveraging technology, and empowering young people to actively participate in shaping the future.

To maximize the potential benefits of psychedelics, we must prioritize responsible research, ensure equitable access, and cultivate a supportive environment that balances scientific exploration with public safety, ultimately leading to a future in which these substances are utilized effectively for therapeutic, spiritual, and innovative purposes. In the same way, to realize the full potential of transhumanism, we must prioritize equitable access to emerging technologies, implement robust regulations, and foster international collaboration to ensure that these advancements benefit all of humanity and do not exacerbate existing social inequalities.

But what I came to realize when I looked at the recommendations of this moonshot manifesto was that I had fallen prey to the same dystopian vision that many around me were spouting. In this view, our future was limited because new technologies and opportunities are the purview of a small community: the uber wealthy; the tech insiders; the policy writers. Yet one additional lesson we can draw from the history of moonshot moments is that innovation is just the beginning of impact, and it is the widespread application of new ideas and technologies that allows humanity to embrace progress.

After all, Gutenberg may have first built that printing press, but he never could have imagined how it would change humanity's conception of space and time once we used it to print newspapers for the masses. Steve Job's iPod would have meant nothing if there had not been so many talented artists producing new music to play from it. Moonshot moments aren't about individuals, but masses.

So, what about the rest of us? Those of us who are neither de' Medicis nor Michelangelos: Are we but passive players in the renaissance of our lifetimes? Hardly. In fact, much of this book has been a siren's song to encourage you to change your own habits and start your own pursuit of a moonshot mindset. Fostering moonshot moments doesn't start at the top: it comes from all of us, putting in daily effort to improve ourselves, increase our knowledge and shared understanding, build empathy and cooperation, and embrace new technologies both ethically and enthusiastically.

The way to start this journey is by first becoming self-aware of your own place in this interconnected, global system of technology, ecology, and society. This means recognizing you are, at this moment, at the beginning of a global, species-evolving renaissance, and you will play a part in this chapter of human destiny. The question becomes how we want to embrace that fortunate position, and what we can do each day to bring about more moonshot moments in our lives, communities, and societies.

Azarian's Meta-Awareness

I want to introduce one last concept on this journey, to draw this discussion of moonshot moments to a close: Bobby Azarian's discussion of meta-awareness. Beyond the familiar realm of self-awareness, Azarian posits a higher level of cognition that recognizes our interconnectedness with a broader system.

He terms this interconnected network the "global brain,"[1] a complex interplay of life-forms and ecosystems that constitute our planet.

Central to Azarian's argument is the ethical imperative inherent in this interconnectedness. Just as we care for our children and their future, we bear a moral obligation to nurture the living network. This entails not only preserving human life but safeguarding all sentient beings and the intricate tapestry of the biosphere. To ensure a future marked by sentience and creativity rather than coldness and lifelessness, we must cultivate a spiritual connection to this grand cosmic scheme.

To achieve this, Azarian proposes meta-awareness as a crucial catalyst. This heightened state of consciousness involves not only being aware of oneself but also understanding one's role within the broader context of the global brain. Meta-awareness is a recognition of interdependence, a realization that our actions cause ripples for the rest of our species and the planet's ecosystem. Azarian emphasizes the need for conscious cultivation of meta-awareness, suggesting that when it becomes widespread, it will propel humanity to unprecedented heights of cooperation and innovation.

Azarian posits that meta-awareness is not a passive state but an active pursuit. It requires conscious cultivation to fully emerge. At its core, meta-awareness is a recursive process: awareness of one's awareness. This self-referential quality allows individuals to step back and examine their thoughts, emotions, and actions in relation to the larger context of the global brain. I like to think of meta-awareness as the recognition of humanity's cosmic destiny, a la Fermi's paradox, and the realization that we must uphold our cosmic responsibility to evolve and progress as a species.

When this heightened level of consciousness becomes widespread, Azarian envisions a cumulative effect. As more individuals develop meta-awareness, the collective consciousness of humanity elevates, fostering a deeper understanding of our interconnectedness and shared destiny. This collective awakening can catalyze innovative solutions to global challenges and inspire unprecedented levels of cooperation.

I wholeheartedly agree with Azarian that "when awareness of awareness of awareness goes mainstream, the resulting synergy will take humanity to new heights."[2] In our increasingly hyper-connected world, we are all more and more aware of the ways in which our morals, actions, and efforts have an impact on the world around us. Acting upon that awareness, and using it to fuel self-improvement, offers the opportunity for anyone to achieve a moonshot mindset in their lives.

Humanity will achieve our full potential when we unlock more moonshot moments to propel us forward. But through meta-awareness, cooperation, and personal growth, we can each be a part of that process. We do not have to wait for Elon Musk or Jensen Huang to solve the problems of our times: rather, we can foster innovation and our species-wide evolution at the individual level, and choose growth over stagnation in our own lives.

Embracing a lifestyle primed for moonshot moments might be easier said than done, but it's definitely not as hard as you would think. What begins as a shift in perspective will snowball—just like my PKM did—into a multidimensional effort that will prepare you to be a part of our larger, global renaissance.

Cultivating a Questioning Mind

Our era sometimes feels like a knowledge paradox: overflowing with information yet starved for depth. The digital age has bestowed upon us an unprecedented wealth of information, accessible with a mere tap. Yet, ironically, this abundance has fostered a culture of intellectual complacency. We've become accustomed to passive consumption, trading the thrill of discovery for the convenience of curated content. The overwhelming influx of data can be paralyzing, stifling curiosity and critical thinking. As a result, I often fear that my generation and those after me have a mindset that assumes it possesses information but lacks the hunger to transform it into understanding.

If meta-awareness is the goal of human consciousness, then curiosity is its indispensable foundation. To ascend to this elevated cognitive state, an individual must perpetually question the nature of reality, self, and the world. They must strive to learn new things everyday—but considering how easy it is to access information, and how many different formats make that information available, growing a curiosity addiction can be more natural than ever before. Curiosity is not merely a cognitive inclination; it is a philosophical stance, a commitment to lifelong inquiry. By cultivating that insatiable desire to understand, one begins to peel back the layers of perception, revealing deeper levels of consciousness.

In this sense, meta-awareness is the culmination of a lifetime of curiosity, a testament to the human spirit's relentless pursuit of knowledge and meaning. It's also something each and every one of us can do to help increase the collective intelligence and ingenuity of humanity.

Curiosity is the ignition key to innovation, not just in labs and hard drives, but in solving the everyday challenges of our lives. By incessantly asking "why," we challenge the status quos around us, dismantling assumptions and revealing hidden opportunities at work, in our families, and in our communities. This inquisitiveness is often met with skepticism, but it's essential to resist the fear of appearing foolish. I've shown that history is replete with examples of ground-breaking ideas that were initially met with ridicule. Embracing uncertainty and questioning the familiar is not merely an intellectual exercise but a courageous act we can integrate into our everyday lives. It's through this relentless pursuit of understanding that we uncover novel solutions, reshape communities and technologies, and drive human progress forward.

The Stoics, Enlightenment scientists, and philosophers of every era were relentless in their pursuit of wisdom. They wrote extensively, from common-place books to manuscripts, to do more than record ideas: it was a process of refinement through the act of articulation. Their work was a dynamic process, a constant questioning and reevaluation. This relentless curiosity and dedication to intellectual growth is a model worthy of emulation. In many ways, this was the exact process by which this book emerged. Curiosity is a process of always adding a little more knowledge, and letting new questions emerge, each time processing and applying what you've learned. By following in their footsteps, we can cultivate a similar spirit of inquiry, leading to deeper understanding and personal growth.

To foster a life of continuous exploration, dedicating specific time for curiosity is essential. Whether it's a morning ritual of listening to a podcast or an evening unwinding with a new book, these dedicated moments allow for unfiltered thought and exploration. Equally important is the pursuit of novel experiences; stepping outside of one's comfort zone ignites new neural pathways and expands perspectives. Failure, often feared, is an invaluable teacher, so try learning new skills as hobbies and finding lessons in their practice.

By learning to embrace the discomfort that can arise from short-term setbacks, individuals can accelerate their growth trajectory in the long term. Ultimately, a commitment to lifelong learning is the cornerstone of a curious mind. It's a journey, not a destination, marked by a ceaseless appetite for knowledge and understanding.

Staying Ahead of the Technology Curve

That coming wave of technological advancement we discussed in Chapter Four can be both exhilarating and intimidating. It's easy to succumb to technophobia, a fear of new technologies. I believe there is an unfortunate strand of public discourse today that has become too dystopic and too fearmongering when it comes to discussions of emerging tech. Naysayers and Luddites alike are all too eager to throw up their hands and declare that everything is burning down around us. But fear is always fought with knowledge. The key to navigating this technological landscape lies, again, in the natural curiosity we all possess.

Beyond just collecting knowledge and being curious, it is important for all of us, at every level of society, to take learning about new tech seriously. Digital literacy is paramount in today's world. Understanding the basics of how technology works is essential for anyone seeking to participate in the digital age. There are books, videos, lessons, and courses available that allow reputable experts to boil down complex technological concepts to digestible bits of learning, meaning that anyone of any background can learn more about AI, synthetic biology, biohacking, psychedelic therapy, and more, in whatever format you prefer, than ever before.

To gain a deeper understanding of emerging technologies, look to the industry leaders. Podcasts, social media, and books offer invaluable insights into the minds of tech pioneers. By following their journeys, you can identify trends, anticipate breakthroughs, and gain inspiration for your own pursuits. The references in the back of this book are a fantastic place to find inspiration: after all, those are the texts that inspired me so deeply already.

Once you've identified technologies that pique your interest, dive deeper. Explore the underlying principles, potential applications, and challenges associated with these innovations. Hands-on experience is invaluable. Experiment with new technologies in your personal life, even if it's just for fun. This hands-on approach fosters a deeper understanding and can help you identify potential use cases.

Building a network of like-minded individuals can accelerate your learning process. Surrounding yourself with technology enthusiasts creates a fertile ground for idea exchange and collaboration. Jump-start your own little pocket of meta-awareness by inspiring learning and creativity in others, and building a community of others who are interested in moonshot ideas and innovations. By sharing knowledge and experiences, you can collectively identify opportunities and overcome challenges.

Remember, the goal isn't to become a tech expert: you're not looking to become Jeff Bezos, you just want to understand a bit about the cutting-edge rocketry being developed at Blue Horizon. Fighting our fears about new technology is about cultivating a mindset of continuous learning and adaptation so that we don't feel like things are hidden or developing faster than we can see. By embracing tech like AI or biohacking to foster a growth mindset, you position yourself and your community to not only survive but thrive in the digital age.

Your Knowledge Vaults Become Your LLM

One step to overcoming the challenge of too much knowledge—the dreaded information overload—is to pair your curiosity with a strong foundation of knowledge management, organized through some sort of system. In Chapter Six,

I introduced the power of a personal knowledge management (PKM) system for organizing notes, ideas, and material, and I cannot express how game-changing having a PKM system has been for me.

By systematically organizing and connecting ideas, PKM gives anyone the ability to transform raw data into a dynamic asset. With a PKM system, you can swiftly retrieve information, spot connections, and foster creativity. This mental agility enhances problem-solving, decision-making, and collaboration. By building a robust knowledge base, you're not just managing information; you're cultivating fertile ground for innovation and growth.

You might be surprised to discover that you're already engaging in PKM practices without realizing it. Those curated Pinterest boards, saved Reels, and carefully constructed YouTube playlists are casual forms of knowledge organization that you're familiar with. You've instinctively grasped the power of associating ideas and information. The next step is to harness this innate ability more consciously. By transitioning from platform-specific collections to a unified PKM system, you gain unprecedented control over your intellectual landscape. Rather than letting algorithms and echo chambers dictate your thought processes, you become the architect of your own data vaults.

By constructing a robust PKM system, you're essentially creating a personalized Large Language Model (LLM) for your life. Just as LLMs learn from vast datasets to generate humanlike text, your PKM becomes a repository of your experiences, knowledge, and insights. As you feed it with information, it grows in sophistication, developing the ability to identify patterns, make connections, and generate new ideas. This personal LLM becomes an invaluable tool for problem-solving, creativity, and decision-making. By investing time in building this intellectual infrastructure, you're empowering yourself with a cognitive advantage.

Insatiable curiosity combined with a well-organized system to interconnect that knowledge can be an unstoppable force. If more of us optimized how we process knowledge, getting the most out of whatever learning we can manage, we as a species will surely see more moonshot-worthy ideas spring up from our collective consciousness. We will also each have the mental flexibility and intellectual engagement to handle the pressing ethical questions that we must all address going forward.

Wielding Your Technomoral Compass

Technomoral virtues are the ethical compass guiding our interactions with technology. They are unique to each individual, requiring a profound process of self-reflection and discovery. While external guidance can offer valuable insights, ultimately, one's technomoral framework must be personally constructed. It is through introspection and careful consideration of our values, beliefs, and aspirations that we can identify the virtues that will shape our technological journey. The true test, however, lies in action. Applying these virtues in our daily lives, from the apps we use to the online communities we participate in, is essential for cultivating a meaningful and ethical relationship with technology.

The ethical fabric of a society is intricately woven from the moral fibers of its individuals. Our personal values serve as the compass guiding our technological decisions, influencing everything from data privacy to artificial intelligence. The choices we make ripple outward, shaping the ethical contours of our collective existence. Public acceptance of new technology—or public outrage against it—can have more impact on development than governmental policy. Critical thinking about our ethical positions empowers us to question the underlying assumptions of technological advancements, to anticipate potential consequences, and to make informed decisions that align with our values. Ultimately, the communal ethical landscape is a reflection of our individual choices, making personal responsibility paramount in shaping a morally sound technological future.

Open and honest conversations about technology are essential for navigating the complex ethical landscape, and those conversations should not only be happening in tech boardrooms, academic conferences, or government offices. By engaging in thoughtful dialogue, we each can create a space to share perspectives, challenge assumptions, and collectively identify potential pitfalls.

Civil discourse is the cornerstone of this process, fostering mutual respect and understanding. It allows us to build consensus around shared values while acknowledging our differences. Moreover, listening to diverse perspectives is crucial for developing comprehensive ethical frameworks. By incorporating a multitude of viewpoints, we can embrace new technologies that are more inclusive, equitable, and sustainable.

Thus far, we've explored the intellectual foundations for cultivating a moonshot mindset: acquiring knowledge, structuring it effectively, and establishing a robust ethical framework. However, intellectual prowess alone is insufficient. To truly unlock the potential for groundbreaking innovation within each and every one of us, we must also optimize our physical and mental well-being. Rather than always feeling a step behind, we must proactively optimize ourselves, thus enhancing critical dimensions of human capacity and preparing ourselves to embrace the challenges and opportunities of the future.

Proactive Mental Health

Moonshot moments often arise from a clear, focused mind. It's no coincidence that many innovators, artists, and creators prioritize mental well-being. When our minds are healthy, we're better equipped to think creatively, solve problems, and take risks. By nurturing our mental health, we cultivate the optimal conditions for innovation to flourish. Just as physical training prepares an athlete for peak performance, mental wellness prepares the mind for its greatest leaps.

Yet too often, we associate genius or artistry with mental struggle. In other cases, we may feel we do not have the time or energy to focus on our mental health. But in our haste to reach for the stars, we often overlook the fertile ground from which such ideas spring: our mental well-being. It's a critical oversight. Mental health isn't a mere precursor to innovation; it is its bedrock. A mind burdened by stress, anxiety, or depression is ill-equipped to cultivate the creativity, resilience, and focus required to nurture a powerful idea into fruition. We must prioritize mental health as an ongoing, proactive pursuit, not as a distant goal to be achieved once the pressure is off. Only then can we create an environment where revolutionary concepts can flourish.

The once mysterious frontier of the human mind today yields its secrets to transhumanist exploration. With tools like neuroimaging and psychedelic therapy, we're peering into the depths of consciousness. This unprecedented access is revealing a profound interconnectedness between mental health, mindset, and physical well-being. Conditions once shrouded in mystery are now understood as intricate biological and psychological processes. As technology continues to evolve, we anticipate groundbreaking discoveries that will revolutionize mental healthcare and our overall understanding of the human experience.

Our ever-advancing understanding of mental health means that we, as individuals, can begin to see how our mental health impacts our daily lives, and work to improve our personal experiences of mental health. But we can also see the effects at a societal level. Despite unprecedented advancements in understanding the human mind, we're grappling with a parallel surge in mental health crises, from drug addiction to depression and suicide. The juxtaposition is stark: never before have we possessed such insights into our mental well-being, yet never before has it been under such strain.

If we can easily see how the mental health crises of individuals become amplified and slow down societal progress as a whole, we can also imagine the opposite: how a society where everyone can access and utilize mental health treatments could be primed and ready for moonshot ideas at every level. Acting on the fact that societal mental health crises are just composites of all the individual mental health crises of its people is itself a form of meta-awareness. That means improving one's mental health will help put an end to societal mental health crises as a whole.

This demands a proactive approach from individuals. Prioritizing mental health, much like physical health, is essential. It's about cultivating resilience, seeking support when needed, and creating a balanced lifestyle. For many of us, it may mean psychedelic therapy under the guidance of a trained professional. In an era marked by constant stimulation and increasing pressures, tending to our mental gardens has become a nonnegotiable for thriving. Do your part to alleviate societal mental health crises while proactively creating a mental environment ripe for audacious thinking,

Winning the Physical Health Game with Transhumanism

I've established the critical role of mental health in preparing for those exhilarating moments of innovation—the moonshots that you will launch within your own life. A sound mind is undoubtedly a potent tool, but it's equally essential to recognize that this mind resides in a physical vessel that requires optimal function. Just as a high-performance vehicle demands meticulous care and maintenance, so too does the human body, the conduit for our thoughts and actions.

Our lives often progress like a roller coaster, with soaring highs of health and wellness interspersed with stressful lows of illness and recovery. As we age, the healthy spans grow shorter and rarer: soon we're chasing down one set of symptoms only to discover some other problem. It's a relentless cycle, a dance with misfortune that interrupts our momentum and drains our energy. Prolonging those vibrant periods of health, maximizing our peak performance, and minimizing the disruptive impact of maladies—this is the ultimate transhumanist goal.

To achieve this, we must transition from a reactive to a proactive approach to health. Instead of simply treating symptoms, we must anticipate and prevent health issues. This is where transhumanism and emerging technologies offer a compelling path forward. We don't have to be billionaire biohackers to reap the benefits of emerging science. By harnessing the power of wearable technology, optimizing nutrition, prioritizing exercise and sleep, and learning more about the frontiers of biohacking, anyone can engineer their bodies to be a little more resilient.

It's time to shift our perspective. Rather than viewing health as a passive state to be maintained, let's embrace it as an active pursuit. When we invest in our physical health, we're making a strategic investment in our individual potential for innovation. A healthy body is the foundation upon which extraordinary ideas are conceived, nurtured, and brought to life.

Embracing a Future-Oriented Moonshot Mindset

As you've made your way through the recommendations of this chapter, the themes we've been grappling with throughout the book—the possibilities of moonshot moments, the ethical implications of technological advancement, the meaning of humanity itself—continue to reverberate. There is no final curtain call for an argument in support of moonshot moments. Technological advancements will continue to surge, reshaping societies and demanding constant ethical reevaluation. This isn't really an epilogue as a place to tie up loose ends; it's the prologue to a grander narrative.

If there is any realization I've made during this journey, it's that our best hope as a species is for all of us to upgrade and optimize ourselves as much as possible: The harder we work to prepare ourselves and our communities for moonshot moments, the more will come, to the benefit of us all. If we were to collectively and intentionally optimize ourselves for moonshot moments, we could unlock a quantum leap in human progress. By aligning our actions with deeply held values and aspirations, we would not only enhance individual fulfillment but also accelerate societal advancement. This heightened level of personal agency could be the catalyst for solving our planet's most pressing challenges—from climate change and resource scarcity to global inequality. Ultimately, by optimizing ourselves, we can optimize our collective future, perhaps even to the point of interstellar exploration.

On a final note, don't just take my word for it: I encourage you to disagree with ideas I've discussed in this book. While I am confident in the ideas presented here, I believe that true intellectual growth stems from that clash of perspectives where we are challenged to see from new viewpoints as a way of understanding our own positions. Many of my favorite writers are futurists I disagree with precisely because their views challenge me to learn more about my own ideas. Knowledge isn't a static entity to be passively consumed; it's a dynamic process shaped through critical engagement. I implore you to approach the concepts of this book—biohacking, hyper-cooperation, artificial intelligence, spirituality in the face of change, cosmic exploration, or psychedelic therapy—with a discerning eye, questioning assumptions and exploring alternative viewpoints. This book has always been intended as a conversation starter, not a definitive statement. May our conversations be plentiful and productive.

Individual transformation is the cornerstone of societal progress. To ensure that our species can fulfill our cosmic destiny through transhumanist evolution, we must begin by understanding ourselves deeply. We must all, as much as we each can, embark on a journey of continuous improvement and empathetic connection to one another. This meta-awareness empowers us to contribute meaningfully to the world. Nurture your curiosity, invest in your personal knowledge management, explore innovative technologies and psychedelic-assisted therapy, and prioritize your physical and mental well-being. Your journey of self-discovery is essential for building a brighter future for all.

My hope is that you won't simply close the cover and move on, but will instead embark on your own intellectual odyssey. I hope I have inspired you to build a personal library filled with the works of Watts, Suleyman, Pollan, Tegmark, Huxley, Harari, and countless others. Each book becomes a key, unlocking new facets of this fascinating conversation, and plants a new seed of curiosity leading to the next rabbit hole. Ideally, this book has left you reassured, dispersing fears about the coming waves of technological and social change. Let this be the springboard that propels you deeper—deeper into research, deeper into discussion, deeper into your own health journey, and deeper into building your own vault of knowledge on these critical issues.

We each have a responsibility to foster the brilliance that lies within and to utilize it to collectively overcome the challenges of our future. Armed with knowledge and a sense of purpose, you now possess the tools to become a catalyst for positive change. The journey toward a future defined by moonshot moments begins with you. Let us approach the future with a sense of optimism and a commitment to building a better world for generations to come.

GLOSSARY FOR
MOONSHOT MOMENTS

Age of Reason (17th & 18th centuries): a European moonshot cluster/golden age that emphasized reason and logic over religious dogma. Thinkers like Newton and Franklin championed critical thinking, scientific inquiry, and observation as the keys to understanding the world.

Agricultural Revolution: estimated to have begun 12,000 years ago, a moonshot moment when humans shifted from nomadic hunter-gathering to settled communities focused on cultivating crops and raising animals. This new, reliable food source allowed populations to grow, leading to permanent settlements and the rise of complex civilizations.

Artificial General Intelligence (AGI): a hypothetical future development in AI. Unlike current AI, AGI refers to a machine capable of human-level intelligence across various cognitive domains. This includes learning, reasoning, adapting, and understanding complex language. AGI is still theoretical, but discussions about its potential impact, both positive and negative, are ongoing.

Artificial Intelligence (AI): a branch of computer science focused on creating intelligent machines. These machines aim to mimic human cognitive abilities like problem-solving and decision-making. AI research includes machine learning, natural language processing, and computer vision. From facial recognition software to self-driving cars, AI is already transforming our lives.

Automation: using machines and computers to replace human labor. While automation increases efficiency and productivity, it also raises concerns about job displacement as some predict many current jobs will be automated. However, others believe automation will create new opportunities requiring skills that complement, rather than compete with, AI and automation.

Axial Age (1000 to 200 BCE): a pivotal period in human history characterized by an extraordinary surge of intellectual and philosophical activity across diverse civilizations. From ancient Greece to China and India, thinkers embarked on a journey of introspection and ethical questioning. This "new awakening of human consciousness" focused on fundamental questions about existence, morality, and the nature of reality. The Axial Age saw the rise of influential figures like Confucius, Buddha, and Socrates, whose ideas continue to shape philosophical and religious thought today.

Ayahuasca: a potent Amazonian tea that has been used for centuries by indigenous people in religious ceremonies. Brewed from plants like Psychotria viridis and Banisteriopsis caapi, it induces powerful visions and altered states of consciousness. Indigenous cultures believe ayahuasca offers a path to healing, divination, and connecting with the spirit world. Ayahuasca use in the West has grown in popularity in recent decades, with retreats offering experiences outside its traditional context. However, its legality varies and its psychoactive properties raise safety concerns.

Biohacking: the practice of experimenting with one's own biology to enhance health, performance, or longevity. Biohackers use various methods, often drawing on technology and scientific advancements. This can range from DIY methods like dietary modifications and exercise routines to more complex interventions like using wearable health trackers or even gene-editing technologies (though the latter is still a controversial and ethically complex area). Biohacking reflects a growing desire on the part of individuals who want to take a more proactive role in managing their own well-being, but it also raises concerns about safety and potential unintended consequences.

Brain-Computer Interface (BCI): a revolutionary technology that bridges the gap between the human brain and a computer. This device allows for direct communication and control without the need for traditional input methods like keyboards or mice. BCIs work by capturing and translating neural signals from the brain, enabling users to interact with computers or even control external devices like prosthetic limbs. BCI technology is still in its early stages of development, but it holds immense promise for applications in medicine, rehabilitation, and potentially even communication and entertainment. However, ethical considerations regarding privacy, security, and potential misuse of BCIs are important areas of discussion as the technology continues to advance.

Climate Change: long-term alterations in a region's or the entire planet's average weather patterns. These shifts can manifest as changes in temperature, precipitation, wind patterns, and extreme weather events. Climate change can occur naturally over long periods, but human activities, primarily the burning of fossil fuels, are accelerating the process at an unprecedented rate. This rapid change disrupts ecosystems, threatens food security, and raises sea levels, impacting coastal communities. Addressing climate change requires global cooperation to slow down decline, along with several moonshot moments to reverse effects.

Coming Wave Technologies: a term coined by Mustafa Suleyman to describe a group of rapidly advancing technologies predicted to have a profound impact on society. These technologies, including artificial intelligence (AI), synthetic biology, robotics, nanotechnology, and quantum computing, are not entirely new, but their rapid development and potential for convergence represent a significant shift. The "coming wave" refers to the idea that these advancements will be inevitable, uncontainable, and will combine to create major transformations in various fields, from healthcare to manufacturing. Understanding and preparing for the potential benefits and challenges posed by these converging technologies is a crucial discussion point for the future.

Commonplace Book: a meticulously crafted notebook or journal used for collecting and organizing information, observations, and ideas. These curated collections often include quotes, excerpts, personal reflections, and connections between various topics. Historically, commonplace books served as a personal knowledge base and a tool for fostering creativity and critical thinking. In the digital age, some adaptations of the commonplace book concept exist in the form of digital note-taking apps and online repositories.

Consciousness: a complex concept in philosophy and neuroscience that refers to the state of being aware of and responsive to one's surroundings, including internal thoughts and feelings. While we each experience consciousness subjectively, its objective nature and underlying mechanisms are still under debate. It is clear that one's sense of consciousness and self are altered while under the influence of psychedelics. Questions surrounding the emergence of consciousness, its relationship to the brain, and the potential for machine consciousness continue to drive scientific exploration and philosophical inquiry.

Cyborg: a hypothetical or existing being that blends biological and mechanical body parts. The term, a portmanteau of "cybernetic" and "organism," often appears in science fiction, where cyborgs may possess enhanced strength, speed, or cognitive abilities due to their prosthetics or implants. While cyborgs remain largely in the realm of fiction, advancements in prosthetics and bioengineering technologies are increasingly blurring the lines between human and machine. Discussions surrounding the ethics and potential implications of human augmentation with technology are becoming increasingly relevant as these fields progress.

Default Mode Network (DMN): a group of interconnected brain regions that become particularly active when our minds are at rest. This is the active part of the brain when you are daydreaming, lost in thought, or simply letting your mind wander. During these moments, the DMN orchestrates processes like introspection, self-reflection, and the construction of our sense of self. This network, first described by Marcus Raichle in 2001, includes key areas like the posterior cingulate cortex, medial prefrontal cortex, and hippocampus. These regions work together to link various brain functions, allowing us to access past memories, ponder the future, and contemplate our place in the world. Neuroimaging studies suggest that the DMN quiets down during psychedelic experiences. This decrease in activity is often linked to the feelings of ego dissolution reported by users, where the rigid boundaries of self seem to soften or even disappear.

Domestication: the long-term process of transforming wild plants and animals into forms that serve human needs. This process involves selective breeding over generations to favor traits beneficial to humans. For example, selective breeding in dogs led to a vast array of breeds with distinct sizes, temperaments, and abilities. Domestication consists of several moonshot moments that have profoundly impacted human history, providing a reliable food source, companionship, and even assisting in labor and transportation.

Ectogenesis: the artificial development of an embryo or fetus entirely outside the human body. This concept, often depicted in science fiction, involves replicating the nurturing environment of the womb within a technological apparatus. While the technology for complete ectogenesis remains in its early stages, advancements in artificial wombs and fetal development techniques raise both ethical considerations and potential benefits. Proponents see ectogenesis as a solution for infertility as well as a necessary development for long-distance space exploration.

Ego Dissolution: a temporary state in which the sense of self weakens or dissolves. This psychological phenomenon can be triggered by various practices like meditation, deep emotional experiences, or the use of certain substances. During ego dissolution, the rigid boundaries we typically construct around our sense of "self" soften, leading to feelings of interconnectedness with everything around us. A common experience for those under the influence of psychedelics, ego dissolution can manifest as a blurring of self and environment, a sense of oneness with the universe, or profound mystical experiences.

Entheogens: from the Greek, "generating the divine within," these are a class of psychoactive substances known for inducing feelings of interconnectedness, a dissolution of ego boundaries, and a deep sense of awe and wonder at the mysteries of existence. These substances, often referred to as psychedelics, can catalyze profound spiritual or mystical experiences. Historically, entheogens have been used for millennia in various cultures by shamans and within religious or spiritual practices. The term itself was coined in the 1970s to distinguish the ancient spiritual applications of these substances from the taboo recreational drug use that emerged in the 1960s.

Extropy: a philosophical movement advocating for the continual improvement of the human condition through the power of applied science and technology.

Extropists believe that technological advancements can be harnessed to overcome limitations like disease, aging, and even scarcity of resources. Their vision includes progress in areas like artificial intelligence, nanotechnology, and cryonics, all aimed at extending human lifespan, health, and cognitive abilities. This movement emphasizes the potential of technology to create a future of abundance and radical human flourishing, and is deeply tied to, though it is not synonymous with, transhumanism.

Fermi Paradox: a philosophical perspective that highlights the apparent contradiction between the vastness of the universe and the lack of concrete evidence for extraterrestrial life. Given the sheer number of stars and potentially habitable planets, many scientists believe the development of intelligent life should be a common occurrence. The paradox lies in the absence of any definitive signs of extraterrestrial civilizations, like interstellar communication or detectable technological signatures. Possible explanations for the paradox range from limitations in our current search methods to the possibility that intelligent life is incredibly rare or self-destructs before reaching advanced stages.

Free Association/Associational Thinking: a technique used in therapy and creative exploration. It involves expressing uncensored thoughts, feelings, and images that arise in the mind, like a stream of consciousness bypassing conscious control. Therapists use it to access the unconscious mind, believing seemingly random connections can reveal underlying patterns, emotions, and memories. By exploring these free associations, therapists gain insights into a client's mental state and potential sources of distress. Similarly, artists and writers can use free association to spark inspiration and generate new ideas by letting their minds wander and connect seemingly unrelated concepts. Research suggests that psychedelics can enhance free association, potentially leading to deeper exploration of the unconscious mind and facilitating both therapeutic breakthroughs and creative insights.

Gamification: the practice of incorporating gamelike elements into nongame contexts to increase user engagement and motivation. These elements can include points, badges, leaderboards, progress bars, and challenges. By tapping into the human desire for competition, achievement, and reward, gamification can make tasks more enjoyable, promote goal setting, and improve data retention.

Golden Age: a metaphor referring to a period of exceptional cultural and intellectual flourishing. Such an era is often characterized by significant advancements in art, literature, science, and philosophy. The Renaissance in Europe is an example, witnessing a surge in artistic expression, scientific discoveries, and humanist thought. The term *golden age* is subjective and can be applied to specific cultures or broader historical periods, so the term "Moonshot Cluster" is also used synonymously in this book.

Having Purpose: the sense of meaning, direction, and significance we derive from our lives. It encompasses our goals, values, and aspirations, and provides a sense of motivation and direction for our actions. Purpose can be found in various aspects of life, such as careers, relationships, creative pursuits, spiritual beliefs, or a dedication to a cause or faith greater than ourselves. The specific source of purpose will vary for each individual, but having it is crucial for well-being and mental health. There are some concerns about challenges to our sense of purpose once AI replaces mundane jobs and tasks.

Hippie Counterculture: a social movement that challenged the dominant cultural norms of materialism, consumerism, and militarism. While the hippie movement's focus on peace and social justice had positive impacts, its association with psychedelics like LSD had unintended consequences. Psychedelic research, which had been gaining traction in the early 1960s as a potential tool for therapy and self-exploration, became entangled with the counterculture's recreational use of LSD. This association, coupled with sensationalized media portrayals and some high-profile incidents, fueled public fear and anxieties. The backlash culminated in President Nixon declaring a "War on Drugs" in 1971, which heavily restricted psychedelic research and hampered scientific understanding of these potentially therapeutic substances for decades.

Hyper-cooperation: a remarkable level of collaboration that transcends simple reciprocity. As discussed by Robert Wright, it's sometimes referred to as a win-win mentality or a non-zero-sum game. This concept involves shared goals, trust, communication, and a willingness to sacrifice individual gain for the collective benefit.

It's considered a crucial factor in human evolutionary success. Unlike other social animals who may cooperate based on immediate benefits like shared food, humans exhibit hyper-cooperation by building long-term social structures, collaborating on complex projects, and even sacrificing for the well-being of the group. This ability to work together effectively toward shared goals has allowed humans to achieve remarkable advancements throughout history, and is the foundation of moonshot clusters.

Interstellar Travel: the hypothetical journey between stars within our galaxy or to those in distant galaxies. Given the vast distances involved, interstellar travel presents immense technological challenges that will require multiple moonshot moments. Current space propulsion technologies would require journeys lasting thousands or even millions of years. However, scientists continue to explore possibilities like nuclear fusion propulsion, antimatter engines, or even theoretical concepts like warp drives that could potentially shorten travel times. While interstellar travel remains in the realm of science fiction for now, ongoing research and technological advancements might pave the way for its realization in the distant future.

Large Language Models (LLMs): a type of artificial intelligence system trained on massive amounts of text data. This allows the system to generate human-quality text, translate languages, write different kinds of creative content, and answer your questions in an informative way. LLMs achieve this by analyzing statistical patterns in the data they are trained on, enabling them to identify relationships between words and concepts. While still under development, LLMs have the potential to revolutionize various fields, from education and communication to content creation and scientific research.

Life 3.0: proposed by physicist Max Tegmark, Life 3.0 refers to a future stage in human existence characterized by the emergence of artificial superintelligence. Tegmark suggests that life has progressed through distinct stages: Life 1.0 representing the origin and evolution of biological organisms, and Life 2.0 encompassing the development of human culture and technology. Life 3.0, then, signifies a future in which AI surpasses human intelligence in all aspects, fundamentally altering society and potentially even our own definition of what it means to be human. Tegmark explores the potential benefits and risks associated with Artificial Superintelligence

(ASI), emphasizing the importance of developing safe and ethical frameworks for its creation and integration into our world.

LSD (lysergic acid diethylamide): also known as acid, LSD is a potent psychedelic compound with powerful mind-altering effects. It was accidentally discovered by Albert Hofmann, a Swiss chemist, in 1938. Initially explored as a potential medication, LSD later gained notoriety for its use in the 1960s counterculture movement. LSD produces a wide range of psychological and sensory effects, including altered perceptions, hallucinations, and intense emotional experiences. While research into the therapeutic potential of LSD was ongoing in the early 1960s, its association with the hippie counterculture and growing public anxieties led to restrictions on its use and research. However, renewed interest has emerged in recent years, with ongoing clinical trials exploring the potential of LSD for treating conditions like anxiety, depression, and addiction.

Machine Learning: a powerful subfield of artificial intelligence that allows computers to learn without explicit programming. Unlike traditional AI approaches that rely on preprogrammed rules, machine learning algorithms can learn from data sets, identifying patterns and improving their performance over time. These algorithms are trained on massive amounts of data, enabling them to recognize complex relationships and make predictions or classifications. Machine learning applications are becoming increasingly widespread, driving advancements in various fields like image and speech recognition, natural language processing, and recommendation systems.

MAVI Method: created by Milan Kordestani, MAVI is a personal knowledge management system designed to streamline information collection and utilization. Mavi stands for Missions, Anchors, Vaults, and Identity, each representing a key component of the system. The Mavi Method emphasizes aligning your PKM system with your broader aspirations, ensuring the information you gather actively supports your goals and personal growth.

Microdosing: the practice of ingesting very small, sub-perceptual doses of psychedelic substances like LSD or psilocybin (the active compound in magic mushrooms) every few days. This practice has gained popularity in recent years, with proponents like James Fadiman, a psychologist and researcher, believing it can enhance mental health and well-being. Microdosing is claimed to promote benefits like increased creativity, focus, mood regulation, and reduced anxiety. While the practice is relatively new and lacks extensive scientific research, anecdotal evidence suggests potential benefits. Several clinical trials are currently underway to investigate the efficacy and safety of microdosing for various mental health conditions.

Mind Uploading: also known as brain downloading, mind uploading is a hypothetical process of transferring a human consciousness to a computer substrate. It involves scanning and replicating the intricate neural connections and information patterns within the brain, essentially creating a digital copy of the mind. This concept remains firmly in the realm of science fiction, as the technology required to perform such a complex and delicate procedure is far beyond current capabilities. Even if mind uploading becomes technically feasible, numerous philosophical and ethical questions remain. While this process may be necessary for space exploration, it is an unlikely path for humanity to tread in the near future.

Moonshot Clusters: a rare phenomenon where multiple moonshot moments, ambitious and revolutionary projects, occur simultaneously. These clusters of innovation can be triggered by various factors, such as significant funding injections, technological breakthroughs in key areas, and the collaborative efforts of brilliant minds. The convergence of these elements creates a period of rapid progress and innovation, with advancements feeding off and accelerating one another. Sometimes referred to as "golden ages," historical examples of moonshot clusters include the Axial Age, the Renaissance, and the Age of Reason. Identifying and fostering conditions that might lead to moonshot clusters is a central theme in discussions around accelerating technological progress.

Moonshot Moments: an ambitious and wildly imaginative project, often associated with groundbreaking technological advancements. These endeavors push the boundaries of what's currently possible, aiming for significant and lasting impacts. Moonshot moments involve setting an immensely ambitious goal, often pushing the boundaries of what's currently achievable. Moonshot moments necessitate creative and outside-the-box approaches to overcome the challenges that stand in the way. They require a willingness to embrace risk and explore untrodden paths in pursuit of potentially transformative results.

Morphological Freedom: the right and ability of individuals to modify their bodies as they see fit. This concept is often discussed in the context of transhumanism and bioethics, exploring the ethical and legal implications of body modifications. Proponents of morphological freedom argue that individuals have the right to autonomy over their bodies, including the freedom to alter their physical form through surgery, implants, or even genetic engineering. This freedom could encompass cosmetic enhancements, performance enhancing modifications, or even radical transformations that blur the lines between human and machine. However, discussions surrounding morphological freedom also raise concerns about potential inequalities, safety risks associated with body modifications, and the ongoing definition of what it means to be human in a world of ever-evolving technology.

Mycelium Network: a vast web of fungal filaments coursing underground exemplifies nature's power of hyper-cooperation. These hidden connectors weave through the soil, forming symbiotic partnerships with plants in a remarkable display of mutual benefit. Mycelium acts as a communication and transportation hub, shuttling water and nutrients between plants, fostering a web of interconnected life. This exchange embodies hyper-cooperation, where individual organisms prioritize the collective good. Mycelium further contributes by decomposing organic matter and returning nutrients to the soil, ensuring a healthy ecosystem. The interconnectedness of these networks inspires research across various fields, with scientists mimicking their information exchange for novel communication technologies and exploring applications in bioremediation and biomaterials.

Nanotechnology: the manipulation of matter on an atomic and molecular scale. This rapidly developing field holds immense potential for creating entirely new materials and devices with unique properties. Scientists can manipulate individual atoms and molecules to engineer materials with specific characteristics, such as exceptional strength, conductivity, or reactivity. Nanotechnology has applications in various fields, including medicine, electronics, energy production, and environmental science. However, nanotechnology also raises concerns about potential environmental and health risks as the behavior of materials at the nanoscale can be quite different from their bulk counterparts.

Nixon's War on Drugs: officially declared in 1971, Nixon's War on Drugs had a profound impact on the research and societal perception of psychedelics. Prior to the 1970s, psychedelics like LSD and psilocybin were being explored for their potential therapeutic applications in treating anxiety, depression, and addiction. However, the rise of recreational psychedelic use in the counterculture movement of the 1960s, coupled with sensationalized media portrayals and growing public anxieties, led to a shift in policy. Psychedelics were classified as Schedule 1 drugs under the Controlled Substances Act, signifying a high potential for abuse and no accepted medical use. This classification severely restricted research on psychedelics, hindering scientific exploration of their potential benefits for decades. However, recent years have witnessed a renewed interest in psychedelic research, with clinical trials investigating their efficacy in treating various mental health conditions.

Non-Zero-Sum Game: a concept from game theory that challenges the traditional win-lose mentality. It expands the possibilities of interaction by highlighting situations where all participants can benefit. Unlike zero-sum games, where one player's gain is another's loss (like a competition with a fixed prize), non-zero-sum games allow for collaborative strategies that create value for everyone involved. The concept of non-zero-sum games is crucial for understanding cooperation and highlights the potential for mutually beneficial outcomes. In a world facing complex challenges, fostering non-zero-sum interactions will be critical for achieving collective progress.

Nootropic: substances believed to enhance cognitive function, improving memory, focus, learning, and overall mental performance. These substances, often referred to as "smart drugs," can be natural supplements, synthetic compounds, or even prescription medications. While some nootropics have shown promise in improving cognitive abilities, research is ongoing, and the long-term effects of many nootropics remain unclear.

PARA Method: developed by productivity expert Tiago Forte, PARA is a personal knowledge management system that categorizes information based on its purpose and relevance in your life. PARA stands for Projects (current goals and action steps), Areas of Interest (broader topics you're curious about), Resources (materials that support your projects and interests), and Archive (information you might need later but isn't currently relevant). This simple framework helps you organize your knowledge base, ensuring you can efficiently capture, categorize, and utilize information to propel your projects, explore your curiosities, and ultimately achieve your goals.

Personal Knowledge Management (PKM): a system or approach for collecting, organizing, reflecting on, and using information to enhance creativity, productivity, and understanding of the world. Information is gathered from various sources, like books, articles, online resources, or even your own thoughts and ideas. This might involve note-taking, digital clipping tools, or mind mapping techniques. The information is then organized in a way that makes it easy to find and retrieve later. PKM is not just about storing information but about creating a system that works for you to capture ideas, resources, and learning in a way that facilitates reflection, analysis, and, ultimately, knowledge creation. Different PKM systems offer various methods for organization, such as tagging, categorizing by project or topic, or using digital folders.

Pillars, Pipelines & Vaults (PPV): created by August Bradley, PPV is a targeted personal knowledge management framework designed for the Notion platform. It goes beyond basic information storage, aiming to transform your PKM setup into a powerhouse for focus, alignment, and knowledge retrieval. PPV focuses on three core elements: Pillars (your high-level goals and aspirations), Pipelines (actionable steps and projects you undertake to achieve your pillars), and Vaults (your central knowledge repository, storing all the resources, notes, and information that fuel your projects and support your goals). By implementing the PPV structure within Notion, users benefit from a clear visualization of their goals, streamlined project management, and an efficient system for capturing and utilizing knowledge.

Prosthetics: artificial replacements for missing body parts. These devices can restore or improve lost physical function and mobility. Prosthetics come in a wide range of forms, from simple hooks for amputees to sophisticated bionic limbs controlled by the user's nervous system. Advancements in materials science and technology have led to the development of increasingly lifelike and functional prosthetics. These transhumanist advancements allow individuals with limb loss or other physical limitations to regain independence, participate in everyday activities, and enjoy a higher quality of life.

Psilocybin Mushrooms: the *Psilocybe* genus, also known as magic mushrooms, are psychedelic substances. Around half of the 200 species in this group contain psilocybin and psilocin, causing their mind-altering effects. Psilocybin is the main psychoactive compound, often used as shorthand for the psilocybin-containing mushrooms. These mushrooms have a rich history of use in religious ceremonies in Central and South America, with their profound effects documented for millennia. Western culture's encounter with psilocybin came in the mid-twentieth century, igniting scientific curiosity about its therapeutic potential and its impact on consciousness. Legal restrictions and concerns about its psychedelic nature hampered research for decades. However, recent times have seen a resurgence in psilocybin research, with promising results in treating various mental health conditions.

Psychedelic Drugs: a category of substances that produce profound alterations in perception, mood, and cognitive processes. These effects can be diverse, ranging from visual and auditory hallucinations to altered states of consciousness, feelings of euphoria, and mystical experiences. Classic psychedelics include LSD, psilocybin, and ayahuasca. While traditionally used in spiritual and religious contexts, psychedelic drugs are now being investigated for their potential therapeutic applications in treating various mental health conditions like depression, anxiety, and addiction. Currently, psychedelics are still classified as a Schedule I drug in the United States. Due to their psychoactive nature, psychedelic drugs should be approached with caution, with a controlled set and setting, and under appropriate guidance.

Quantum Supremacy: the ability of a quantum computer to perform a specific calculation that would be impossible, or take an unreasonably long time, for even the most powerful classical computers. Classical computers rely on bits, which can be either 0 or 1. Quantum computers, however, utilize qubits, which can exist in a superposition of both states simultaneously. This unique property allows quantum computers to explore a vast number of possibilities concurrently, tackling certain problems exponentially faster than classical machines. Achieving quantum supremacy is a significant milestone in the development of quantum computing, demonstrating its potential to revolutionize fields like materials science, drug discovery, and cryptography. While quantum computers are still in their early stages, the possibilities unlocked by quantum supremacy hold immense promise for future technological advancements.

Rare Earth Elements (REEs): a group of seventeen chemical elements found on the periodic table with similar chemical properties. Despite the name, REEs are not particularly rare in the Earth's crust, but they are often dispersed and difficult to extract in large quantities. However, REEs play a critical role in various modern technologies, including smartphones, electric vehicle batteries, lasers, and wind turbines. The increasing demand for these technologies has intensified focus on sustainable and ethical REE mining practices as well as space exploration and asteroid mining.

Regenerative Medicine: a rapidly evolving transhumanist field focused on repairing or replacing damaged tissues and organs. This field utilizes various strategies, including stem cell therapy, tissue engineering, and gene editing, to promote the body's natural healing processes or create entirely new tissues and organs. Regenerative medicine holds immense potential for treating a wide range of diseases and injuries, including heart disease, diabetes, spinal cord injuries, and burns. While research is still ongoing, advancements in regenerative medicine offer hope for improved healthcare outcomes and potentially life-altering treatments for individuals with debilitating conditions.

Renaissance (14th–17th centuries): a golden age of profound cultural and intellectual transformation in Europe. This era witnessed a flourishing of creativity, a renewed interest in classical Greek and Roman thought, and a shift toward reason and logic in explaining the world. The Renaissance saw the rise of great artists like Leonardo da Vinci, Michelangelo, and Raphael, the emergence of groundbreaking scientific discoveries by Galileo Galilei, and the flourishing of literary giants like William Shakespeare and Miguel de Cervantes. The ultimate moonshot cluster, the Renaissance's legacy continues to shape our understanding of art, science, philosophy, and human potential.

Reskilling: the process of acquiring new skills and knowledge to adapt to the ever-evolving demands of the job market. The rapid pace of technological change can render certain skills obsolete, necessitating continuous learning and adaptation from the workforce. Reskilling can involve formal education programs, online courses, professional development workshops, or even self-directed learning initiatives. In the face of artificial intelligence's growing presence, reskilling empowers individuals to develop complementary skills that work alongside AI, ensuring their continued value and employability in the future of work.

Salience Network: a network in the brain that acts as a critical filter when we experience information overload. This network, located primarily in the anterior cingulate cortex and insula, prioritizes incoming stimuli, determining what's most relevant to the current situation. Imagine it as a conductor in an orchestra, directing your attention to the instruments playing the most important parts of the music. The Salience Network plays a crucial role in various functions, including decision-making, emotional processing, and maintaining focus. Psychedelics can weaken the network's filtering ability, leading to a heightened awareness of internal and external stimuli. By quieting the Salience Network's filtering, psychedelics may allow access to normally suppressed thoughts, emotions, and memories. This unfiltered state could potentially underlie the therapeutic benefits some people experience with psychedelics, such as increased self-awareness, emotional processing, and a sense of interconnectedness.

Schedule I Substances: a classification within the Controlled Substances Act of the United States encompassing drugs with a high potential for abuse and no currently accepted medical use. Heroin, cocaine, LSD, and psilocybin mushrooms are some examples of Schedule I drugs. The classification is supposed to be based on scientific evidence regarding a drug's potential for addiction, dependence, and its relative safety for medical use. However, the Schedule I designation can be controversial, as some argue that it hinders research on potentially beneficial applications of these substances.

Self-quantification: the transhumanist practice of tracking and analyzing personal data using various technologies and tools. This data can include sleep patterns, activity levels, heart rate, mood, and even dietary choices. By tracking and analyzing these metrics, individuals can gain valuable insights into their overall health, identify areas for improvement, and set personalized goals for self-improvement. Self-quantification tools range from simple fitness trackers to smartphone apps and advanced wearable devices, empowering individuals to become active participants in managing their health.

Set and Setting: a concept used in psychedelic therapy that refers to the mental and emotional state of the person (set) and the physical and social environment (setting) in which the psychedelic experience takes place. Factors like the location, presence of a trusted guide, and overall atmosphere can significantly impact the nature of the psychedelic journey. Both set and setting play a crucial role in shaping the experience and maximizing its therapeutic potential. The terms are usually credited to Timothy Leary.

Singularity: a hypothetical moment in the future when technological advancement accelerates beyond human comprehension or control. Proponents of the singularity theory believe that artificial intelligence will eventually reach a point of superintelligence, surpassing human cognitive abilities in all domains. This rapid AI development could lead to profound and unpredictable changes in society, potentially transforming everything from the nature of work to the very fabric of human existence. The concept of the singularity remains highly speculative, with experts holding varying views on its likelihood and potential consequences.

Superintelligence: a hypothetical level of intelligence that far surpasses human capabilities. While human intelligence is based on biological limitations, superintelligence could exist in artificial forms, exceeding human cognitive abilities in areas like processing speed, memory capacity, and problem-solving. The concept of superintelligence raises a number of ethical and philosophical questions that will require global ethical debate. While superintelligence remains a theoretical possibility, ongoing advancements in AI research necessitate careful consideration of the ethical implications associated with such powerful technology.

Surveillance: the close observation of individuals or groups by others. While historically used for security purposes, concerns around surveillance have grown immensely in the digital age. Authoritarian regimes are increasingly leveraging advanced technologies like artificial intelligence (AI) and robotics to monitor their citizens on a massive scale. This can include facial recognition systems, social media monitoring, and even drone surveillance. These practices raise significant ethical concerns regarding privacy, freedom of expression, and the potential for social control.

Synthetic Biology: a revolutionary transhumanist field within biotechnology. It goes beyond traditional genetic engineering, aiming to design and build entirely new biological systems from scratch. This field utilizes tools like DNA synthesis, gene editing, and metabolic engineering to create novel organisms with specific functionalities. Synthetic biology holds immense potential for various applications, including developing new drugs and therapies, engineering biofuels, and even creating sustainable materials.

Technogoals: specific, measurable objectives that define the desired outcomes of technological development. They reflect our shared values and priorities, ensuring that technological innovation serves humanity's best interests. For instance, a technogoal might be to develop AI systems that are fair, unbiased, and transparent, or to create clean energy technologies that address climate change. By establishing clear technogoals, we can guide technological advancements toward a more just and sustainable future.

Technomoral Virtues: ethics that guide our interactions with technology and its consequences. These virtues, first conceptualized by Susan Vallor, extend beyond technical expertise and encompass essential human values like humility, justice, and care. By cultivating these technomoral virtues, we can ensure that technology is used for good and fosters a more ethical and humane future.

Time Blocking: a productivity technique that involves dividing your calendar into specific blocks of time dedicated to particular tasks to regain control of your schedule. This method allows you to focus your attention on one task at a time, minimizing distractions and maximizing your efficiency. Time blocking helps users prioritize tasks, visualize their workday, and ensure users allocate enough time for what truly matters.

Transhumanism: a philosophical and scientific movement that advocates for the transformation of the human condition by leveraging technological and scientific advancements. Proponents of transhumanism believe that we can overcome our current biological limitations through fields like bioengineering, artificial intelligence, and cognitive science. This could involve enhancing human lifespans, augmenting our physical and cognitive abilities, and even achieving a state of mind-body fusion with technology. Transhumanism raises a number of ethical questions regarding the potential misuse of technology, the definition of what it means to be human, and the social implications of a society with enhanced humans. While the long-term goals of transhumanism may seem futuristic, advancements in areas like gene editing and prosthetics are already blurring the lines between human and machine.

Win-Win Scenario: a situation where all parties involved in an interaction benefit and achieve a positive outcome. This concept is often used in negotiation and conflict resolution, where the goal is to find mutually beneficial solutions. From an evolutionary perspective, win-win scenarios can be advantageous, as cooperation can lead to greater survival and reproductive success for all parties involved. Hyper-cooperation, a concept explored in fields like evolutionary biology and game theory by writers like Robert Wright and Joseph Henrich, refers to situations where individuals cooperate beyond what might be expected based on self-interest alone, potentially leading to win-win outcomes on a larger scale. The concept of win-win scenarios is not only relevant in individual interactions but also holds significance in areas like international relations and environmental sustainability, where finding solutions that benefit all stakeholders is crucial.

Zettelkasten: a German word translating to "notebox," a Zettelkasten system utilizes note cards or digital equivalents to capture atomic notes. These atomic notes are small, self-contained pieces of information, often consisting of a single idea, observation, or quote. The key concept of Zettelkasten lies in the linking of these atomic notes thematically. By creating connections between notes based on ideas and concepts, users can build a rich and interconnected web of knowledge. The Zettelkasten system encourages active engagement with information, promoting critical thinking and fostering a deeper understanding of complex topics. While traditionally implemented with physical note cards, digital adaptations of the Zettelkasten method have emerged, allowing for a more portable and searchable knowledge base.

WORKS CITED

Chapter 1: Moonshots as a Framework for the Future

1. Wright, R. (2000). *Nonzero: The Logic of Human Destiny*. Vintage. 332.
2. Watts, A. (2013). *The Joyous Cosmology: Adventures in the Chemistry of Consciousness*. New World Library. 42–44.
3. More, Max. (2013). A Letter to Mother Nature. In M. More & N. Vita-More (Eds.), *The Transhumanist Reader: Classical and Contemporary Essays on the Science, Technology, and Philosophy of the Human Future* (pp. 449–451). Wiley-Blackwell. 449–450.

Chapter 2: Humanity's History of Audacious Thinking

1. Tumulty, K. (2019, June 23). The Moonshot Mindset Once Came from the Government. No Longer. *Wall Street Journal*. https://www.wsj.com/articles/the-moonshot-mind-set-once-came-from-the-government-no-longer-11563152820
2. Davies, Alex (2019, July 20). Apollo 11's Successful Moonshot, 21st Century Style. *Wired*. https://www.wired.com/story/apollo-11-moonshot-21st -century/
3. Selected Reading from Kierkegaard. (n.d.). In *Introduction to Philosophy*. Oklahoma State University Library. Retrieved from https://open.library .okstate.edu/introphilosophy/chapter/selected-reading-from-kierkegaard/
4. Baraniuk, Chris. (2015, February 26). Ancient Customer Feedback Technology Lasts Millennia. *New Scientist*. Retrieved from https://www.newscientist.com/article/dn27063-ancient-customer-feedback-technology-lasts-millennia/

5. Foster, Benjamin R. (1974.) Humor and Cuneiform Literature. *Journal of the Ancient Near Eastern Society, 6* (1).

6. Wright, R. (2000). *Nonzero: The Logic of Human Destiny.* Vintage. 93–100.

7. Henrich, J. (2016). *The Secret of Our Success: How Culture Is Driving Human Evolution, Domesticating Our Species, and Making Us Smarter.* Princeton University Press. 260–264.

8. Jaspers, Karl. 1953. *The Origin and Goal of History,* Bullock, Michael (Tr.). Routledge.

9. Harari, Y. N. (2015). *Sapiens: A Brief History of Humankind.* Harper Perennial. 249–256.

10. Armstrong, Karen. (2006). *The Great Transformation: The Beginning of Our Religious Traditions* (1st ed.). Knopf.

11. Smith, Andrew. (2015). Between Facts and Myth: Karl Jaspers and the Actuality of the Axial Age. *International Journal of Philosophy and Theology, 76*(4), 315–334. https://www.tandfonline.com/doi/full/10.1080/21692327.2015.1136794

12. Spinney, L. (2019, December 9). When Did Societies Become Modern? "Big History" Dashes Popular Idea of Axial Age. *Nature.* https://www.nature.com/articles/d41586-019-03785-w

13. Mullins, D. A., Hoyer, D., Collins, C., Currie, T., Feeney, K., François, P., Savage, P. E., Whitehouse, H., & Turchin, P. (2018). A Systematic Assessment of "Axial Age" Proposals Using Global Comparative Historical Evidence. *American Sociological Review, 83*(3), 596–626. https://doi.org/10.1177/0003122418772567

14. Bekhrad, J. (2017, April 6). The Obscure Religion That Shaped the West. *BBC.* Retrieved from https://www.bbc.com/culture/article/20170406-this-obscure-religion-shaped-the-west

15. Harari, Y. N. (2015). *Sapiens: A Brief History of Humankind.* Harper Perennial. 246–247.

16. Armstrong, K. (1994). *A History of God: The 4,000-Year Quest of Judaism, Christianity and Islam.* Ballantine Books.

17. Harari, Y. N. (2015). *Sapiens: A Brief History of Humankind.* Harper Perennial. 249–253.

18. Naydler, J. (2019). *The Shadow of the Machine: The Hidden History of Consciousness.* Psyche Books. 32–42.

19. Harari, Y. N. (2015). *Sapiens: A Brief History of Humankind.* Harper Perennial. 219.

20. Harari, Y. N. (2015). *Sapiens: A Brief History of Humankind.* Harper Perennial. 221.

21. Hajar, R. M. D. (2013). The Air of History (Part V) Ibn Sina (Avicenna): The Great Physician and Philosopher. *Heart Views, 14*(4), 196–201. https://www.ncbi.nlm.nih.gov/pmc/articles/PMC3621228/

22. Hajar, R. M. D. (2013). The Air of History (Part III): The Golden Age in Arab Islamic Medicine an Introduction. *Heart Views, 14*(1), 43–46. https://www.ncbi.nlm.nih.gov/pmc/articles/PMC3621228/

23. Hajar, R. M. D. (2013). The Air of History (Part IV): Great Muslim Physicians Al Rhazes. *Heart Views, 14*(2), 93–95. https://www.ncbi.nlm.nih.gov/pmc/articles/PMC3752886/

24. Hajar, R. M. D. (2013). The Air of History (Part V) Ibn Sina (Avicenna): The Great Physician and Philosopher. *Heart Views, 14*(4), 196–201. https://www.ncbi.nlm.nih.gov/pmc/articles/PMC3970379/

25. Harari, Y. N. (2015). *Sapiens: A Brief History of Humankind.* Harper Perennial. 256–263.

26. Naydler, J. (2019). *The Shadow of the Machine: The Hidden History of Consciousness.* Psyche Books. 133–142.

27. Naydler, J. (2019). *The Shadow of the Machine: The Hidden History of Consciousness.* Psyche Books. 160–168.

28. Harari, Y. N. (2015). *Sapiens: A Brief History of Humankind.* Harper Perennial. 295.

29. Henrich, J. (2016). *The Secret of Our Success: How Culture Is Driving Human Evolution, Domesticating Our Species, and Making Us Smarter.* Princeton University Press.

Chapter 3: Humanity's Future (Choose Wisely)

[There are no citations in this fictional chapter.]

Chapter 4: The Inevitable Wave of Challenges

1. Suleyman, M., & Bhaskar, M. (2023). *The Coming Wave: Technology, Power, and the Twenty-First Century's Greatest Dilemma.* Crown Publishing. 7.

2. Dartmouth News. (n.d.). Artificial Intelligence (AI) Coined at Dartmouth. *Dartmouth News*. Retrieved from https://home.dartmouth.edu/about/artificial-intelligence-ai-coined-dartmouth

3. Samuel, A. L. (1959). Some Studies in Machine Learning Using the Game of Checkers. *IBM Journal, 3*(3), 535–554. Retrieved from https://www.cs.virginia.edu/~evans/greatworks/samuel1959.pdf

4. Hendler, James. (2008). Avoiding Another AI Winter. *IEEE Software, 25*(2), 2–7. Retrieved from https://web.archive.org/web/20120212012656/http://csdl2.computer.org/comp/mags/ex/2008/02/mex2008020002.pdf

5. Krizhevsky, A., Sutskever, I., & Hinton, G. E. (2012). ImageNet Classification with Deep Convolutional Neural Networks. In *NIPS'12: Proceedings of the 25th International Conference on Neural Information Processing Systems—Volume 1* (pp. 1097–1105).

6. Suleyman, M., & Bhaskar, M. (2023). *The Coming Wave: Technology, Power, and the Twenty-First Century's Greatest Dilemma*. Crown Publishing. 118.

7. Suleyman, M., & Bhaskar, M. (2023). *The Coming Wave: Technology, Power, and the Twenty-First Century's Greatest Dilemma*. Crown Publishing. 120.

8. Webster, G., Creemers, R., Kania, E., & Triolo, P. (2017). Full Translation: China's New Generation Artificial Intelligence Development Plan. *DigiChina*. Retrieved from https://digichina.stanford.edu/work/full-translation-chinas-new-generation-artificial-intelligence-development-plan-2017/

9. Pilipiszyn, Ashley. (2021). GPT-3 Powers the Next Generation of Apps. *OpenAI Blog*. Retrieved from https://openai.com/blog/gpt-3-apps

10. Future of Life Institute. (2023). An Open Letter: Pause Giant AI Experiments. *Future of Life Institute*. Retrieved from https://futureoflife.org/open-letter/pause-giant-ai-experiments/

11. Morange, M. (2008). *A History of Molecular Biology (Vol. 3)*. Cold Spring Harbor Laboratory Press. 319.

12. Hood, L., & Rowen, L. (2013). The Human Genome Project: Big Science Transforms Biology and Medicine. *Genome Medicine, 5*, 79. https://doi.org/10.1186/gm483

13. Denby, C. M., Li, R. A., Vu, V. T., et al. (2018). Industrial Brewing Yeast Engineered for the Production of Primary Flavor Determinants in Hopped Beer. *Nature Communications, 9*(1), 965. https://doi.org/10.1038/s41467-018-03293-x

14. Zheng, Q., Ning, R., Zhang, M., & Deng, X. (2023). Biofuel Production as a Promising Way to Utilize Microalgae Biomass Derived from Wastewater: Progress, Technical Barriers, and Potential Solutions. *Frontiers in Bioengineering and Biotechnology, 11*, 1250407. https://doi.org/10.3389/fbioe.2023.1250407

15. Schooley, R. T., Biswas, B., Gill, J. J., et al. (2017). Development and Use of Personalized Bacteriophage-Based Therapeutic Cocktails to Treat a Patient with a Disseminated Resistant Acinetobacter Baumannii Infection. *Antimicrobial Agents and Chemotherapy, 61*(10), e00954–17. https://doi.org/10.1128/aac.00954-17

16. Economist Staff. (2023, April 13). The Human Genome Project Transformed Biology. *Economist.* Retrieved from https://www.economist.com/leaders/2023/04/13/the-human-genome-project-transformed-biology

17. Sovijärvi, O., Arina, T., & Halmetoja, J. (2018). *Biohacker's Handbook,* L. Viitaniemi (Ed.). Biohacker Center BHC Inc.

18. Suleyman, M., & Bhaskar, M. (2023). *The Coming Wave: Technology, Power, and the Twenty-First Century's Greatest Dilemma.* Crown Publishing. 82.

19. Naydler, J. (2019). *The Shadow of the Machine: The Hidden History of Consciousness.* Psyche Books. 199–205.

20. Darrach, Brad. (1970, November 20). Meet Shaky, the First Electronic Person. *LIFE.* Retrieved from https://books.google.com/books?id=2FMEAAAAMBAJ&pg=PA56-IA4#v=onepage&q&f=false

21. Ackerman, E. (2022). Even as It Retires, ASIMO Still Manages to Impress. *IEEE Spectrum.* Retrieved from https://spectrum.ieee.org/honda-asimo

22. Boston Dynamics. (2019). Spot Launch [Video]. YouTube., https://www.youtube.com/watch?v=wlkCQXHEgjA.

23. Ho, C., Tsakonas, E., Tran, K., Cimon, K., Severn, M., Mierzwinski-Urban, M., Corcos, J., & Pautler, S. (2012). Robot-Assisted Surgery Compared with Open Surgery and Laparoscopic Surgery. *CADTH Technology Overviews, 2*(2), e2203.

24. Kim, H., Lee, J., Heo, U., Jayashankar, et al. (2024). Skin Preparation–Free, Stretchable Microneedle Adhesive Patches for Reliable Electrophysiological Sensing and Exoskeleton Robot Control. *Science Advances, 10*(3).

25. Ajayan, P. M., Stephan, O., Colliex, C., & Trauth, D. (1994). Aligned Carbon Nanotube Arrays Formed by Cutting a Polymer Resin—Nanotube Composite. *Science, 265* (5176), 1212–1214.

26. Zhang, J., Wilkinson, D. P., Wang, H. & Liu, Z.-S. (2005). FTIR and Electro-
    chemical Observation of Water Content Reduction in a Thin Nafion® Film
    Induced by an Impregnation of Metal Complex Cations. *Electrochimica Acta,
    50*(20), 4082–4088.

27. Prabhu, S., Kim, H., Wang, J., Wang, J., & Gu, Z. (2021). Recent Advances in
    Nanocomposite-Based Biosensors for Disease Diagnosis. *Nano Today, 41,*
    101112.

28. Ahmed, S., Yasuda, N., & Kawahara, N. (2015). A Review on the Recent Prog-
    ress in the Preparation and Applications of Superhydrophobic Surfaces Using
    Polymer and Ceramic Nanoparticle Composite Materials. *Composites Commu-
    nications, 1*(4), 70–84.

29. Kurzweil, R. (2006). *The Singularity Is Near: When Humans Transcend Biology.*
    Penguin Books.

30. Oberdörster, G., Elder, A., & Risch, H. (2005). Nano Objects and Nanoparti-
    cles in Medicine: Risk and Benefit. *Nanomedicine: Nanotechnology, Biology, and
    Medicine, 1*(1), 33–44.

31. Martínez, G., Merinero, M., Pérez-Aranda, M., Pérez-Soriano, E. M., Ortiz, T.,
    Begines, B., & Alcudia, A., (2020). Environmental Impact of Nanoparticles' Ap-
    plication as an Emerging Technology: A Review. *Materials (Basel), 14*(1), 166.

32. National Academies of Sciences, Engineering, and Medicine. (2019).
    Quantum Computing: Progress and Prospects, E. Grumbling & M. Horowitz
    (Eds.). Retrieved from https://nap.nationalacademies.org/catalog/25196
    /quantum-computing-progress-and-prospects

33. Aaronson, Scott. (2019). Google's Quantum Breakthrough. *New York Times.* Re-
    trieved from https://www.nytimes.com/2019/10/30/opinion/google-quantum
    -computer-sycamore.html

34. Swayne, Matt. (2024). Quantinuum: DeepMind Scientists Use AI to
    Minimize Tricky T Gates in Step Toward Practical Quantum Com-
    puting. *The Quantum Insider.* Retrieved from https://thequantuminsider
    .com/2024/02/28/quantinuum-deepmind-scientists-use-ai-to-minimize
    -tricky-t-gates-in-step-toward-practical-quantum-computing/

35. Rahimi, M., & Asadi, F. (2023). Oncological Applications of Quantum Machine
    Learning. *Technology in Cancer Research & Treatment, 22,* 15330338231215214.
    https://doi.org/10.1177/15330338231215214

36. Letzter, Rafi. (2023). How Boeing Is Using Quantum Computing. *IBM Research Blog*. Retrieved from https://research.ibm.com/blog/boeing -case-study

37. Soutar, C., Buchholz, S., Dannemiller, D. & Bhuta, M. (2023). Industry Spending on Quantum Computing Will Rise Dramatically. Will It Pay Off? *Deloitte Insights*. Retrieved from https://www2.deloitte.com/xe/en/insights /industry/financial-services/financial-services-industry-predictions/2023 /quantum-computing-in-finance.html

38. National Academies of Sciences, Engineering, and Medicine. (2019). *Quantum Computing: Progress and Prospects*, E. Grumbling & M. Horowitz (Eds.). Retrieved from https://nap.nationalacademies.org/catalog/25196/ quantum-computing-progress-and-prospects

39. Chakravarty, S., New, M., Dasgupta, P., Dessai, S., Patt, A., Shaw, M., & Verghese, G., (2007). Sharing Global Greenhouse Gas Emissions Fairly: A Study on Sharing Rights and Mitigation Burdens. *World Development*, *36*(2), 191–219.

40. David, Attenborough. (2020). David Attenborough: A Life on Our Planet. *Netflix*. https://www.netflix.com/title/80216393.

41. England, T. (2021). The Rise of Authoritarianism: Crescendo or Inflection Point? *MEA and Associates*. Retrieved from https://www.meandahq.com/ the-rise-of-authoritarianism-crescendo-or-inflection-point/

42. Freedom House. (2023). *Freedom in the World 2023*. Retrieved from https:/ /freedomhouse.org/sites/default/files/2023-03/FIW_World_2023 _DigtalPDF.pdf

43. World Economic Forum. (2020). *The Future of Jobs Report 2020*. Retrieved from https://www.weforum.org/publications/the-future-of-jobs-report -2020/

44. World Economic Forum. (2023). *The Future of Jobs Report 2023: Digest*. Retrieved from https://www.weforum.org/publications/the-future-of-jobs-report-2023 /digest/

45. McKinsey Global Institute. (2017). *Jobs Lost, Jobs Gained: Workforce Transitions in a Time of Automation*. Retrieved from https://www.mckinsey.com/ ~/media/mckinsey/industries/public%20and%20social%20sector/our %20insights/what%20the%20future%20of%20work%20will%20mean%20for %20jobs%20skills%20and%20wages/mgi-jobs-lost-jobs-gained-executive -summary-december-6-2017.pdf

46. Bostrom, N. (2024). *Deep Utopia: Life and Meaning in a Solved World*. Ideapress Publishing. 146.

47. Bostrom, N. (2024). *Deep Utopia: Life and Meaning in a Solved World*. Ideapress Publishing. 147.

48. Bostrom, N. (2024). *Deep Utopia: Life and Meaning in a Solved World*. Ideapress Publishing. 151–162.

49. Suleyman, M., & Bhaskar, M. (2023). *The Coming Wave: Technology, Power, and the Twenty-First Century's Greatest Dilemma*. Crown Publishing. 218.

50. Bostrom, N. (2019). The Vulnerable World Hypothesis. *Global Policy, 10*(4), 455–476. Retrieved from https://nickbostrom.com/papers/vulnerable.pdf

51. Harris, S. (2020). *Making Sense: Conversations on Consciousness, Morality, and the Future of Humanity*. Ecco Publishing. 326.

52. Kurpjuhn, T. (2021). The Guide to Ransomware: How Businesses Can Manage the Evolving Threat. *Computer Fraud & Security, 2019*(11), 14-16.

53. Suleyman, M. & Bhaskar, M. (2023). *The Coming Wave: Technology, Power, and the Twenty-First Century's Greatest Dilemma*. Crown Publishing. 160.

54. Suleyman, M., & Bhaskar, M. (2023). *The Coming Wave: Technology, Power, and the Twenty-First Century's Greatest Dilemma*. Crown Publishing. 161.

55. Williams, G. (2017). WannaCry and International Law in Cyberspace. *Just Security*. Retrieved from https://www.justsecurity.org/50038/wannacry-international-law-cyberspace/

56. SysArc. (2017). 5 Information Security Lessons from the Global Petya Cyberattack. *SysArc*. Retrieved from https://www.sysarc.com/cyber-security/5-information-security-lessons-global-petya-cyberattack/

57. Greenberg, A. (2018). The Untold Story of NotPetya, the Most Devastating Cyberattack in History. *Wired*. Retrieved from https://www.wired.com/story/notpetya-cyberattack-ukraine-russia-code-crashed-the-world/

58. Greenberg, A. (2018). The Untold Story of NotPetya, the Most Devastating Cyberattack in History. *Wired*. Retrieved from https://www.wired.com/story/notpetya-cyberattack-ukraine-russia-code-crashed-the-world/

59. Cimpanu, C. (2021). Ransomware Gang Tries to Extort Apple Hours Ahead of Spring Loaded Event. *The Record*. Retrieved from https://therecord.media/ransomware-gang-tries-to-extort-apple-hours-ahead-of-spring-loaded-event

60. BBC. (2021). Meat Giant JBS Pays $11m in Ransom to Resolve Cyber-attack. *BBC News*. Retrieved from https://www.bbc.com/news/business-57423008

61. Kerner, S.M., (2022). Colonial Pipeline Hack Explained: Everything You Need to Know. *TechTarget*. Retrieved from https://www.techtarget.com/whatis /feature/Colonial-Pipeline-hack-explained-Everything-you-need-to-know

62. Process Unity. (2021). REvil's Reign: Kaseya VSA Ransomware Supply Chain Attack Decoded. *ProcessUnity*. Retrieved from https://www.processunity .com/revils-reign-kaseya-vsa-ransomware-supply-chain-attack-decoded/

63. BBC. (2022). REvil Ransomware Gang Arrested in Russia. *BBC News*. Retrieved from https://www.bbc.com/news/technology-59998925

64. Tucker, J. B. (2001). *War of the Bugs: Biological Weapons and the New Security Threat*. W. W. Norton & Company.

65. Murakami, Y. (2006). The Tokyo Subway Sarin Attack: A Case Study of Technological Terrorism. *Terrorism and Political Violence, 18*(3), 497–514.

66. Lifton, R. J. (2000). *Destroying the World to Save It: Aum Shinrikyo, Apocalyptic Violence, and the New Global Religion*. Metropolitan Books.

67. Murakami, Y. (2006). The Tokyo Subway Sarin Attack: A Case Study of Technological Terrorism. *Terrorism and Political Violence, 18*(3), 497–514.

68. Kobori, I. (2005). Medical Response to the Tokyo Subway Sarin Attack: 10-Year Follow-Up Study. *Journal of Urban Health, 82*(2), 309–319.

69. Kobori, I. (2005). Medical Response to the Tokyo Subway Sarin Attack: 10-Year Follow-Up Study. *Journal of Urban Health, 82*(2), 309–319.

Chapter 5: Humanity and Hyper-Cooperation

1. Vallor, S. (2016). *Technology and the Virtues: A Philosophical Guide to a Future Worth Wanting*. Oxford University Press.

2. Wright, R. (2000). *Nonzero: The Logic of Human Destiny*. Vintage.

3. Richerson, P. J., & Boyd, R. (2005). *Not By Genes Alone: How Culture Transformed Human Evolution*. University of Chicago Press.

4. Tomasello, M. (2021). *Becoming Human: A Theory of Ontogeny*. Harvard University Press.

5. Kuipers, B. (2022). Trust and Cooperation. *Frontiers in Robotics and AI, 9*. https://doi.org/10.3389/frobt.2022.676767

6. Wright, R. (2000). *Nonzero: The Logic of Human Destiny*. Vintage. 6–7.

7. Tegmark, M. (2017). *Life 3.0: Being Human in the Age of Artificial Intelligence.* Vintage Books. 25.

8. Martin, W. F., Garg, S., Zimorski, V. (2015). Endosymbiotic Theories for Eukaryote Origin. *Philos Trans R Soc Lond B Biol Sci*, 370(1678), 20140330. https://doi.org/10.1098/rstb.2014.0330

9. Main, D. (2020). World's Oldest Fungi, Found in Fossils, May Rewrite Earth's Early History. *National Geographic Magazine.* Retrieved from https://www.nationalgeographic.com/science/articleoldest-fungus-fossils-found-earth-history

10. Gamillo, E. (2022). Mushrooms May Communicate with Each Other Using Electrical Impulses. *Smithsonian Magazine.* Retrieved from https://www.smithsonianmag.com/smart-news/mushrooms-may-communicate-with-each-other-using-electrical-impulses-180979889/

11. Gorzelak, M. A., Asay, A. K., Pickles, B. J., & Simard, S. W. (2015). Inter-Plant Communication Through Mycorrhizal Networks Mediates Complex Adaptive Behaviour in Plant Communities. *AoB Plants, 7*, plv050. https://doi.org/10.1093/aobpla/plv050

12. Gorzelak, M. A., Asay, A. K., Pickles, B. J., & Simard, S. W. (2015). Inter-Plant Communication Through Mycorrhizal Networks Mediates Complex Adaptive Behaviour in Plant Communities. *AoB Plants, 7*, plv050. https://doi.org/10.1093/aobpla/plv050

13. Ton, L. L. M., Jones, N. S. & Boddy, L. (2017). The Mycelium as a Network. *Microbiology Spectrum, 5.* https://doi.org/10.1128/microbiolspec.funk-0033-2017

14. Dykes, J. (2016). Mycoremediation: The Under-Utilised Art of Fungi Clean-Ups. *Geographical.* Retrieved from https://geographical.co.uk/science-environment/mycoremediation-the-art-of-fungi-clean-ups

15. Vaksmaa, A., Guerrero-Cruz, S., Ghosh, P., Zeghal, E., Hernando-Morales, V., & Niemann, H. (2023). Role of Fungi in Bioremediation of Emerging Pollutants. *Frontiers in Marine Science, 10.* https://doi.org/10.3389/fmars.2023.1070905

16. Van der Kooi, C. J., & Ollerton, J. (2020). The Origins of Flowering Plants and Pollinators. *Science, 368*(6497), 1306–1308. https://doi.org/10.1126/science.aay3662

17. Van der Kooi, C. J., & Ollerton, J. (2020). The Origins of Flowering Plants and Pollinators. *Science, 368*(6497), 1306–1308. https://doi.org/10.1126/science.aay3662

18. Gowlett, J. A. (2016). The Discovery of Fire by Humans: A Long and Convoluted Process. *Philosophical Transactions of the Royal Society B: Biological Sciences, 371*(1696), 20150164. https://doi.org/10.1098/rstb.2015.0164

19. Gowlett, J. A. (2016). The Discovery of Fire by Humans: A Long and Convoluted Process. *Philosophical Transactions of the Royal Society B: Biological Sciences, 371*(1696), 20150164. https://doi.org/10.1098/rstb.2015.0164

20. Harari, Y. N. (2015). *Sapiens: A Brief History of Humankind.* Harper Perennial. 13–14.

21. Gowlett, J. A. (2016). The Discovery of Fire by Humans: A Long and Convoluted Process. *Philosophical Transactions of the Royal Society B: Biological Sciences, 371*(1696), 20150164. https://doi.org/10.1098/rstb.2015.0164

22. Aiello, L. C., & Wheeler, P. (1995). The Expensive-Tissue Hypothesis: The Brain and the Digestive System in Human and Primate Evolution. *Current Anthropology,* 36(2), 199–221.

23. Henrich, J. (2016). *The Secret of Our Success: How Culture is Driving Human Evolution, Domesticating Our Species, and Making Us Smarter.* Princeton University Press.

24. Packer, C., & Ruttan, L. (1988). The Evolution of Cooperative Hunting. *The American Naturalist,* 132(2).

25. Henrich, J. (2016). *The Secret of Our Success: How Culture is Driving Human Evolution, Domesticating Our Species, and Making Us Smarter.* Princeton University Press. 256–257.

26. Harari, Y. N. (2015). *Sapiens: A Brief History of Humankind.* Harper Perennial. 63–65.

27. D'Huy, J. (2016). Scientists Trace Society's Myths to Primordial Origins. *Scientific American.* Retrieved from https://www.scientificamerican.com/article/scientists-trace-society-rsquo-s-myths-to-primordial-origins

28. Campbell, J. (1988). *The Power of Myth.* Anchor Press.

29. Harari, Y. N. (2015). *Sapiens: A Brief History of Humankind.* Harper Perennial. 30.

30. Henrich, J. (2016). *The Secret of Our Success: How Culture Is Driving Human Evolution, Domesticating Our Species, and Making Us Smarter.* Princeton University Press.

31. Harari, Y. N. (2015). *Sapiens: A Brief History of Humankind.* Harper Perennial. 112–114.

32. Henrich, J. (2016). *The Secret of Our Success: How Culture Is Driving Human*

33. *Evolution, Domesticating Our Species, and Making Us Smarter*. Princeton University Press. 92–93.

34. Tegmark, M. (2017). *Life 3.0: Being Human in the Age of Artificial Intelligence*. Vintage Books.

35. Kurzweil, R. (2006). *The Singularity Is Near: When Humans Transcend Biology*. Penguin Books.

36. Bostrom, N. (2014). *Superintelligence: Paths, Dangers, Strategies*. Oxford University Press.

37. Suleyman, M., & Bhaskar, M. (2023). *The Coming Wave: Technology, Power, and the Twenty-First Century's Greatest Dilemma*. Crown Publishing. 74.

38. Bostrom, N. (2014). *Superintelligence: Paths, Dangers, Strategies*. Oxford University Press.

39. Tegmark, M. (2017). *Life 3.0: Being Human in the Age of Artificial Intelligence*. Vintage Books. 260.

40. United Nations Department of Economic and Social Affairs (2024). Sustainable Development Goals: The 17 Goals. *United Nations*. Retrieved from https://sdgs.un.org/goals

41. Vallor, S. (2016). *Technology and the Virtues: A Philosophical Guide to a Future Worth Wanting*. Oxford University Press.

42. Tegmark, M. (2017). *Life 3.0: Being Human in the Age of Artificial Intelligence*. Vintage Books. 249.

43. Tegmark, M. (2017). *Life 3.0: Being Human in the Age of Artificial Intelligence*. Vintage Books. 260.

44. Suleyman, M., & Bhaskar, M. (2023). *The Coming Wave: Technology, Power, and the Twenty-First Century's Greatest Dilemma*. Crown Publishing. 278.

45. Suleyman, M., & Bhaskar, M. (2023). *The Coming Wave: Technology, Power, and the Twenty-first Century's Greatest Dilemma*. Crown Publishing. 259.

Chapter 6: Associational Thinking and Personal Knowledge Management

1. Forte, T. (2022). *Building a Second Brain: A Proven Method to Organize Your Digital Life and Unlock Your Creative Potential*. Atria Books. 33–38.

2. Forte, T. (2022). *Building a Second Brain: A Proven Method to Organize Your Digital Life and Unlock Your Creative Potential*. Atria Books. 19–22.

3. Darnton, R. (2009). *The Case for Books: Past, Present, and Future.* Public Affairs. 224.

4. Kukil, K. (2011). Woolf in the World: A Pen and a Press of Her Own. *Smith College Libraries.* Retrieved from https://www.smith.edu/libraries/libs /rarebook/exhibitions/penandpress/

5. Houghton Library, Harvard University. (2024). Catalog search results for Goethe, Johann Wolfgang von, 1749–1832. Curiosity. Retrieved from https://curiosity.lib.harvard.edu/reading/catalog?f%5Bcreators-contributors _ssim%5D%5B%5D=Goethe%2C+Johann+Wolfgang+von%2C+1749-1832&f %5Brepository_ssim%5D%5B%5D=Houghton+Library%2C+Harvard +University

6. Forte, T. (2022). *Building a Second Brain: A Proven Method to Organize Your Digital Life and Unlock Your Creative Potential.* Atria Books. 145–149.

7. Moss, A. (1996). *Printed Commonplace-Books and the Structuring of Renaissance Thought.* Oxford University Press.

8. Bacon, F. (1883). *The Promus of Formularies and Elegancies.* Longman, Greens and Company.

9. Benjamin Franklin Commonplace-Book. (1718–1720). *Colonial Society of Massachusetts Transactions, 10* (1904–1906), 191–225.

10. Franklin, B. (1959 ed.) *The Papers of Benjamin Franklin, January 6, 1706 Through December 31, 1734,* Leonard W. Labaree (Ed.). Yale University Press. 3–4.

11. Dolbear, S. (2019). John Locke's Method for Common-Place Books (1685). *Public Domain Review.* Retrieved from https://publicdomainreview.org /collection/john-lockes-method-for-common-place-books-1685/

12. Soft White Underbelly. (2024). A Talk on Romantic Relationships with Orion Taraban [Video]. YouTube. https://www.youtube.comwatch?v=ryCnP-nBibs

13. Forte, T. (2022). *Building a Second Brain: A Proven Method to Organize Your Digital Life and Unlock Your Creative Potential.* Atria Books. 90–95.

14. Forte, T. (2022). *Building a Second Brain: A Proven Method to Organize Your Digital Life and Unlock Your Creative Potential.* Atria Books. 91–92.

15. Ahrens, S. (2022). *How to Take Smart Notes: One Simple Technique to Boost Writing, Learning and Thinking.* Independently Published.

16. Ahrens, S. (2022). *How to Take Smart Notes: One Simple Technique to Boost Writing, Learning and Thinking.* Independently Published.

17. Zettelkasten. (2024). Overview. Retrieved from https://zettelkasten.de /overview/

18. Forte, T. (2022). *Building a Second Brain: A Proven Method to Organize Your Digital Life and Unlock Your Creative Potential.* Atria Books. 231–237.

19. Bradley, A. (2020). Notion Design with Systems Thinking Approach [Video]. YouTube. https://www.youtube.com/watch?v=CxscGZwk0S4

20. Bradley, A. (2020). Intro & Overview of Pillars, Pipelines & Vaults—Notion Life OS [Video]. YouTube. https://www.youtube.com/watch?v =d93SGaf82OM

21. Forte, T. (2022). *Building a Second Brain: A Proven Method to Organize Your Digital Life and Unlock Your Creative Potential.* Atria Books. 238–239.

22. Forte, T. (2022). *Building a Second Brain: A Proven Method to Organize Your Digital Life and Unlock Your Creative Potential.* Atria Books. 39.

23. Giaro, M. (2024). How to Use Obsidian as a Zettlekasten: The Ultimate Tutorial. Retrieved from https://mattgiaro.com/obsidian-zettelkasten/

24. Thacker, T. F. (2023). Logseq: A Powerful Tool for Thought. *Medium.* Retrieved from https://tfthacker.medium.com/logseq-a-powerful-tool-for -thought-9058dec80dbe

25. Abdaal, A. (n.d.). Trying Out Roam Research [Newsletter]. Retrieved from https://aliabdaal.com/newsletter/trying-out-roam-research/

26. Gupta, A. (2023). Top 16 Digital Tools That Every Researcher Should Know About. *Researcher Life.* Retrieved from https://researcher.life/blog/article /top-digital-tools-for-researchers/

27. Matthijs. (2021). Roam, Bear, or Notion: Which of These Popular Apps Works for You? *Medium.* https://medium.com/a-journey-towards-better -notes/roam-bear-or-notion-which-of-these-popular-apps-works-for-you -12fe984ddb7d

28. Bru, T. (2020). How to Create an Exothermic Note-Taking System. *Medium.* https://medium.com/age-of-awareness/my-exothermic-note-taking-system -432aec80d5b8

29. Yuan, X. (2022). Evidence of the Spacing Effect and Influences on Perceptions of Learning and Science Curricula. *Cureus, 14*(1), e21201. https://doi .org/10.7759/cureus.21201

30. Forte, T. (2022). *Building a Second Brain: A Proven Method to Organize Your Digital Life and Unlock Your Creative Potential.* Atria Books. 198–217.

31. Seavers, L. (2021). *Time-Blocking: Your Method to Supercharge Productivity & Reach Your Goals.* Independently published.

32. Abdaal, A. (2023). *Feel-Good Productivity: How to Do More of What Matters to You.* Celadon Books. 131–133.

33. Abdaal, A. (2023). *Feel-Good Productivity: How to Do More of What Matters to You.* Celadon Books. 30–31.

34. Seavers, L. (2021). Productivity Tips: Time Blocking [Video]. YouTube. https://www.youtube.com/watch?v=65e2qScV_K8

35. Watson, S. (2021). Dopamine: The Pathway to Pleasure. *Harvard Health Publishing.* Retrieved from https://www.health.harvard.edu/mind-and-mood/dopamine-the-pathway-to-pleasure

36. Rich, V. (2024). Fitness App Gamification: A Trend You Cannot Miss. *Shakuro.* Retrieved from https://shakuro.com/blog/fitness-app-gamification-in-2021-a-trend-you-cant-miss

37. Abdaal, A. (2023). *Feel-Good Productivity: How to Do More of What Matters to You.* Celadon Books.

38. Abdaal, A. (2023). *Feel-Good Productivity: How to Do More of What Matters to You.* Celadon Books.

Chapter 7: We Need to Change Our Minds: A Case for Psychedelics

1. McDonald, C. (2019). Intolerable Genius: Berkeley's Most Controversial Nobel Laureate. *California Magazine.* Retrieved from https://alumni.berkeley.edu/california-magazine/winter-2019/intolerable-genius-berkeleys-most-controversial-nobel-laureate

2. Schoch, R. (1994). Q&A—a Conversation with Kerry Mullis. *California Monthly, 105*(1), 20.

3. McDonald, C. (2019). Intolerable Genius: Berkeley's Most Controversial Nobel Laureate. *California Magazine.* Retrieved from https://alumni.berkeley.edu/california-magazine/winter-2019/intolerable-genius-berkeleys-most-controversial-nobel-laureate

4. Mullis, K. B. (1993). The Polymerase Chain Reaction. Nobel Lecture. Retrieved from https://www.nobelprize.org/prizes/chemistry/1993/mullis/lecture/

5. McDonald, C. (2019). Intolerable Genius: Berkeley's Most Controversial Nobel Laureate. *California Magazine.* Retrieved from https://alumni.berkeley.edu/california-magazine/winter-2019/intolerable-genius-berkeleys-most-controversial-nobel-laureate

6. Uthaug M. V., Lancelotta, R., van Oorsouw, K., Kuypers, K. P. C., Mason, N., Rak, J., et al. (2019). A Single Inhalation of Vapor from Dried Toad Secretion Containing 5-methoxy-N,N-dimethyltryptamine (5-MeO-DMT) in a Naturalistic Setting is Related to Sustained Enhancement of Satisfaction with Life, Mindfulness-Related Capacities, and a Decrement of Psychopathological Symptoms. *Psychopharmacology (Berl). 236*(9):2653–2666.

7. Davis A. K., Xin Y., Sepeda N., & Averill L.A. (2023). Open-Label Study of Consecutive Ibogaine and 5-MeO-DMT Assisted-Therapy for Trauma-Exposed Male Special Operations Forces Veterans: Prospective Data from a Clinical Program in Mexico. *Am J Drug Alcohol Abuse. 49*(5):587-596.

8. Riaz K., Suneel S., Hamza Bin Abdul Malik M., Kashif T., Ullah I, Waris A, et al. (2023). MDMA-Based Psychotherapy in Treatment-Resistant Post-Traumatic Stress Disorder (PTSD): A Brief Narrative Overview of Current Evidence. *Diseases, 11*(4):159.

9. Letheby, C. (2021). *Philosophy of Psychedelics.* Oxford University Press. 64–66.

10. Nutt, D. (2022). *Drugs Without the Hot Air: Making Sense of Legal and Illegal Drugs.* UIT Cambridge Ltd. 254.

11. Pollan, M. (2019). *How to Change Your Mind: What the New Science of Psychedelics Teaches Us About Consciousness, Dying, Addiction, Depression, and Transcendence.* Penguin Press. 301.

12. Pollan, M. (2019). *How to Change Your Mind: What the New Science of Psychedelics Teaches Us About Consciousness, Dying, Addiction, Depression, and Transcendence.* Penguin Press. 304.

13. De Gregorio D., Enns J. P., Nuñez N. A., Posa L., Gobbi G. (2018). d-Lysergic Acid Diethylamide, Psilocybin, and Other Classic Hallucinogens: Mechanism of Action and Potential Therapeutic Applications in Mood Disorders. *Prog. Brain Res. 242,* 69–96.

14. Preller K. H., Razi A., Zeidman P., Stämpfli P., Friston K. J., Vollenweider F. X. (2019). Effective Connectivity Changes in LSD-Induced Altered States of Consciousness in Humans. *PNAS 116,* 2743–2748.

15. Madsen M. K., Stenbæk D. S., Arvidsson A., Armand S., Marstrand-Joergensen M. R., Johansen S. S., et al. (2021). Psilocybin-Induced Changes in Brain Network Integrity and Segregation Correlate with Plasma Psilocin Level and Psychedelic Experience. *Eur. Neuropsychopharmacol. 50,* 121–132.

16. Pollan, M. (2019). *How to Change Your Mind: What the New Science of Psyche-delics Teaches Us About Consciousness, Dying, Addiction, Depression, and Tran-scendence.* Penguin Press. 315.

17. Letheby, C. (2021). *Philosophy of Psychedelics.* Oxford University Press. 99.

18. Fox, K. C., Nijeboer, S., Dixon, M. L., Floman, J. L., Ellamil, M., Rumak, S. P., Sedlmeier, P., & Christoff, K. (2014). Is Meditation Associated with Altered Brain Structure? A Systematic Review and Meta-Analysis of Morphometric Neuroimaging in Meditation Practitioners. *Neuroscience & Biobehavioral Reviews, 43,* 48–73.

19. Fox, K. C., Dixon, M. L., Nijeboer, S., Girn, M., Floman, J. L., Lifshitz, M., El-lamil, M., Sedlmeier, P., & Christoff, K. (2016). Functional Neuroanatomy of Meditation: A Review and Meta-Analysis of 78 Functional Neuroimaging Investigations. *Neuroscience & Biobehavioral Reviews, 65,* 208–228.

20. Letheby, C. (2021). *Philosophy of Psychedelics.* Oxford University Press. 98–101.

21. Lebedev, A. V., Lövdén, M., Rosenthal, G., Feilding, A., Nutt, D. J., & Car-hart-Harris, R. L. (2015). Finding the Self by Losing the Self: Neural Correlates of Ego-Dissolution under Psilocybin. *Human Brain Mapping, 36*(8), 3137–3153.

22. Carhart-Harris, R. L. & Nutt, D. J. (2017). Serotonin and Brain Function: A Tale of Two Receptors. *Journal of Psychopharmacology, 31*(9), 1091–1120. https://doi.org/10.1177/0269881117725915

23. Reissig, C. J., Eckler, J. R., Rabin, R. A., & Winter, J. C. (2005). The 5-HT1A Receptor and the Stimulus Effects of LSD in the Rat. *Psychopharmacology, 182*(2), 197–204. https://doi.org/10.1007/s00213-005-0068-6

24. Erkizia-Santamaría, I., Alles-Pascual, R., Horrillo, I., Meana, J. J. & Ortega, J. E. (2022). Serotonin 5-HT2A, 5-HT2c and 5-HT1A Receptor Involvement in the Acute Effects of Psilocybin in Mice: In Vitro Pharmacological Profile and Modulation of Thermoregulation and Head-Twitch Response. *Biomedicine & Pharmacotherapy, 154,* 113612. https://doi.org/10.1016/j.biopha.2022.113612

25. López-Giménez, J. F. & González-Maeso, J. (2018). Hallucinogens and Serotonin 5-HT2A Receptor-Mediated Signaling Pathways. *Current Topics in Behavioral Neurosciences, 36,* 45–73. https://doi.org/10.1007/7854_2017_478

26. Vargas, M. V., Dunlap, L. E., Dong, C., Carter, S. J., Tombari, R. J., Jami, S. A., Cameron, L. P., Patel, S. D., Hennessey, J. J. & Olson, D. E. (2023). Psychedelics

Promote Neuroplasticity through the Activation of Intracellular 5-HT2A Receptors. *Science, 379*(6633), 700–706. https://doi.org/10.1126/science.adf0435

27. Tang, Y. Y., Hölzel, B. K. & Posner, M. I. (2015). The Neuroscience of Mindfulness Meditation. *Frontiers in Human Neuroscience,* 8, Article 20. https://doi.org/10.3389/fnhum.2014.00020

28. Pollan, M. (2019). *How to Change Your Mind: What the New Science of Psychedelics Teaches Us About Consciousness, Dying, Addiction, Depression, and Transcendence.* Penguin Press. 320.

29. Letheby, C. (2021). *Philosophy of Psychedelics.* Oxford University Press. 184–191.

30. Richards, W. (2015). *Sacred Knowledge: Psychedelics and Religious Experiences.* Columbia University Press.

31. Davis, J. & Lampert, J. (Rockingstone Group). (2022). Expediting Psychedelic-Assisted Therapy Adoption in Clinical Settings. In H. McCormack, L. Raines, H. Harbin, J. Glastra & C. Gross (Eds.), *BrainFutures.*Retrieved from https://www.brainfutures.org/mental-health-treatment/expeditingpatadoption/

32. Richards, W. (2015). *Sacred Knowledge: Psychedelics and Religious Experiences.* Columbia University Press. 129–31.

33. Carod-Artal, F. J. (2015). Hallucinogenic Drugs in Pre-Columbian Mesoamerican Cultures. *Neurología* (English ed.), *30*(1), 42–49.

34. Akers B. P., Ruíz J. F., Piper A., Ruck C. A. P. (2011). A Prehistoric Mural in Spain Depicting Neurotropic Psilocybe Mushrooms? *Econ. Bot.* 65. 121–128.

35. Samorini, G. (2020). Mushroom Effigies in Archaeology: A Methodological Approach. In K. Feeney (Ed.), *Fly Agaric: A Compendium of History, Pharmacology, Mythology, and Exploration* (pp. 269–296). Fly Agaric Press.

36. Hedau, V. N. & Anjankar, A. P. (2022). Psychedelics: Their Limited Understanding and Future in the Treatment of Chronic Pain. *Cureus, 14*(8), e28413. https://doi.org/10.7759/cureus.28413

37. Van Court, R. C., Wiseman, M. S., Meyer, K. W., Ballhorn, D. J., Amses, K. R., Slot, J. C., Dentinger, B. T. M., Garibay-Orijel, R. & Uehling, J. K. (2022). Diversity, Biology, and History of Psilocybin-Containing Fungi: Suggestions for Research and Technological Development. *Fungal Biology, 126*(4), 308–319.

38. Romero, O. (2022). Mazatec Shamanism and Psilocybin Mushrooms. *Chacruna.* Retrieved from https://chacruna.net/mazatec-shamanism-and-psilocybin-mushrooms/

39. McKenna, T. (1993). *Food of the Gods: The Search for the Original Tree of Knowledge—a Radical History of Plants, Drugs, and Human Evolution.* Bantam Books.

40. Sessa, B. (2019). *The Psychedelic Renaissance: Reassessing the Role of Psychedelic Drugs in 21st Century Psychiatry and Society.* 2nd ed. Aeon Academic. 175.

41. Allegro, J. M. (2009). *The Sacred Mushroom and The Cross: A Study of the Nature and Origins of Christianity Within the Fertility Cults of the Ancient Near East.* Doubleday.

42. Carod-Artal, F. J. (2015). Hallucinogenic Drugs in Pre-Columbian Mesoamerican Cultures. *Neurología* (English ed.), *30*(1), 42–49.

43. Wasson, R. G. (1957). Seeking the Magic Mushroom. *Life Magazine,* 42–61.

44. Gregoire, C. (2020). Inside the Movement to Decolonize Psychedelic Pharma. *proto.life Magazine.* Retrieved from https://proto.life/2020/10/inside -the-movement-to-decolonize-psychedelic-pharma/

45. Letcher, A. (2006). *Shroom: A Cultural History of the Magic Mushroom.* Faber and Faber. 97–98.

46. Sessa, B. (2019). *The Psychedelic Renaissance: Reassessing the Role of Psychedelic Drugs in 21st Century Psychiatry and Society.* 2nd ed. Aeon Academic. 183–184.

47. Sessa, B. (2019). *The Psychedelic Renaissance: Reassessing the Role of Psychedelic Drugs in 21st Century Psychiatry and Society.* 2nd ed. Aeon Academic. 173–178.

48. Sessa, B. (2019). *The Psychedelic Renaissance: Reassessing the Role of Psychedelic Drugs in 21st Century Psychiatry and Society.* 2nd ed. Aeon Academic. 72–73.

49. Lucchesi, E. (2023). This Hallucinogenic Fungus Might Be Behind the Salem Witch Trials. *Discover Magazine.* Retrieved from https://www.discovermagazine.com/ mind/this-hallucinogenic-fungus-might-be -behind-the-salem-witch-trials

50. Sessa, B. (2019). *The Psychedelic Renaissance: Reassessing the Role of Psychedelic Drugs in 21st Century Psychiatry and Society.* 2nd ed. Aeon Academic. 74–75.

51. Sessa, B. (2019). *The Psychedelic Renaissance: Reassessing the Role of Psychedelic Drugs in 21st Century Psychiatry and Society.* 2nd ed. Aeon Academic. 135.

52. Drug Enforcement Administration. (2024). Controlled Substances Act (CSA). Retrieved from https://www.dea.gov/drug-information/csa

53. Sessa, B. (2019). *The Psychedelic Renaissance: Reassessing the Role of Psychedelic Drugs in 21st Century Psychiatry and Society.* 2nd ed. Aeon Academic. 129–130.

54. Pollan, M. (2019). *How to Change Your Mind: What the New Science of Psychedelics Teaches Us About Consciousness, Dying, Addiction, Depression, and Transcendence.* Penguin Press. 172.

55. Pollan, M. (2019). *How to Change Your Mind: What the New Science of Psychedelics Teaches Us About Consciousness, Dying, Addiction, Depression, and Transcendence.* Penguin Press. 313.

56. Nour, M. M., Evans, L., & Carhart-Harris, R. L. (2017). Psychedelics, Personality and Political Perspectives. *Journal of Psychoactive Drugs, 49*(3), 182–191.

57. Markoff, J. (2006). What the Dormouse Said: How the Sixties Counterculture Shaped the Personal Computer Industry. Penguin Publishing Group.

58. Wainwright, O. (2017). Designers on Acid: the Tripping Californians who Paved the Way to Our Touchscreen World. *The Guardian.* https://www.theguardian.com/artanddesign/2017/may/11/design-museum-california-designing-freedom-tech-design

59. Stamets, P. (2008). 6 Ways Mushrooms Can Save the World. *TED.* Retrieved from https://www.ted.com/speakers/paul_stamets

60. Fadiman, J. (2011). *The Psychedelic Explorer's Guide: Safe, Therapeutic, and Sacred Journeys.* Park Street Press.

61. Robinson, M. (2015). When Silicon Valley Takes LSD. *CNN.* Retrieved from https://money.cnn.com/2015/01/25/technology/lsd-psychedelics-silicon-valley/

62. Markoff, J. (2006). *What the Dormouse Said: How the Sixties Counterculture Shaped the Personal Computer Industry.* Penguin Publishing Group.

63. Markoff, J. (2006). *What the Dormouse Said: How the Sixties Counterculture Shaped the Personal Computer Industry.* Penguin Publishing Group.

64. Isaacson, W. (2021). *Steve Jobs.* Simon & Schuster.

65. Shead, S. (2020). Peter Thiel Backs Berlin Start-Up Making Psychedelics in $125 million Round. *CNBC.* https://www.cnbc.com/2020/11/23/peter-thiel-backs-psychedelics-startup-atai.html

66. Meyer, D. (2022). A 36-Year-Old CEO Saw How Psychedelics Treated His Best Friend's Mental Illness. Now His Peter Thiel-Funded Firm Wants to Bring the Drugs to the Masses. *Fortune.* https://fortune.com/2022/12/05/florian-brand-atai-life-sciences-psychedelics-mental-illness/

67. Henning, E. (2024). Thiel-Backed Psychedelic Firm Atai said to Buy Stake in Beckley. *Bloomberg.* https://www.bloomberg.com/news/articles/2024-01-04/thiel-backed-psychedelic-firm-atai-said-to-buy-stake-in-beckley

68. Grind, K., Glazer, E., Elliott, R., & Jones, C. (2024). The Money and Drugs That Tie Elon Musk to Some Tesla Directors. *The Wall Street Journal*. https://www.wsj.com/tech/elon-musk-tesla-money-drugs-board -61af9ac4?mod=hp_lead_pos7

69. Ng, K. (2023) Elon Musk 'Microdoses Ketamine to Manage Depression', Report Says. *The Independent*. https://www.independent.co.uk/life-style/health-and-familieselon-musk-ketamine-depression-microdose-b2365648.html

70. Musk, E. [@elonmusk]. (2022, April 24). *I have serious concerns about SSRIs, as they tend to zombify people. Occasional use of Ketamine is a much better option, in my opinion. I have a prescription for when my brain chemistry sometimes goes super negative.* [Post]. X. https://twitter.com/elonmusk /status/1687663413877714944

71. Grind, K. & Bindley, K. (2023). Magic Mushrooms. LSD. Ketamine. The Drugs That Power Silicon Valley. *The Wall Street Journal*. https://www.wsj.com/articles/silicon-valley-microdosing-ketamine-lsd-magic-mushrooms -d381e214

72. Berger, C. (2024) Sam Altman says taking psychedelics 'significantly changed' his mindset. *Fortune*. https://fortune.com/2024/09/25/sam -altman-psychedelic -experience-openai-ceo/

73. Berger, C. (2024) Sam Altman says taking psychedelics 'significantly changed' his mindset. *Fortune*. https://fortune.com/2024/09/25 /sam-altman-psychedelic-experience-openai-ceo/

74. Ghosh, S. (2023). Elon Musk, Sergey Brin Reportedly Taking Ketamine, Magic Mushroom, Tesla CEO Says, 'Better Option than . . . *Mint*. https://www.livemint.com/news/world/elon-musk-sergey-brin-reportedly|-taking-ketamine -magic-mushroom-tesla-ceo-says-better-option-than-11688006452319.html

75. Pollan, M. (2019). *How to Change Your Mind: What the New Science of Psychedelics Teaches Us About Consciousness, Dying, Addiction, Depression, and Transcendence*. Penguin Press. 301–304.

76. Carhart-Harris, R. L., Erritzoe, D., Williams, T., et al. (2012). Neural Correlates of the Psychedelic State as Determined by fMRI Studies with Psilocybin. *Proceedings of the National Academy of Sciences of the United States of America, 109*(6), 2138–2143.

77. Robinson, R. (2024) AI Is Rapidly Changing Psychedelic Medicine, New Report Shows. *Double Blind*. https://doubleblindmag.com/ai -is-changing-psychedelic-medicine/

78. Sarris J., Halman A., Urokohara A., Lehrner M., & Perkins, D. (2012). Artificial Intelligence and Psychedelic Medicine. *Annals of the New York Academia of Sciences, 540*(1), 5-12.

79. Sarris J., Halman A., Urokohara A., Lehrner M., & Perkins, D. (2012). Artificial Intelligence and Psychedelic Medicine. *Annals of the New York Academia of Sciences, 540*(1), 5-12.

80. Cherian, K. N., Keynan, J. N., Anker, L. et al. (2024). Magnesium–Ibogaine Therapy in Veterans with Traumatic Brain Injuries. *Nat Med* 30, 373–381.

81. Pollan, M. (2019). *How to Change Your Mind: What the New Science of Psychedelics Teaches Us About Consciousness, Dying, Addiction, Depression, and Transcendence.* Penguin Press. 190.

82. Sessa, B. (2019). *The Psychedelic Renaissance: Reassessing the Role of Psychedelic Drugs in 21st Century Psychiatry and Society.* 2nd ed. Aeon Academic. 30–31.

83. Fadiman, J. & BigThink (2024). Try Psychedelics. Access Transcendence. [Video]. YouTube. https://www.youtube.com/watch?v=xFRZuh0f9Ys

84. Davis, A. K., Barrett, F. S., May, D. G., et al. (2021). Effects of Psilocybin-Assisted Therapy on Major Depressive Disorder: A Randomized Clinical Trial. *JAMA Psychiatry, 78*(5), 481–489. https://doi.org/10.1001/jamapsychiatry .2020.3285

85. Griffiths, R. R., Johnson, M. W., Carducci, M. A., Umbricht, A., Richards, W. A., Richards, B. D., Cosimano, M. P. & Klinedinst, M. A. (2016). Psilocybin Produces Substantial and Sustained Decreases in Depression and Anxiety in Patients with Life-Threatening Cancer: A Randomized Double-Blind Trial. *Journal of Psychopharmacology, 30*(12), 1181–1197. https://doi.org/10.1177/0269881116675513

86. Moreno, F. A., Wiegand, C. B., Taitano, E. K. & Delgado, P. L. (2006). Safety, Tolerability, and Efficacy of Psilocybin in 9 Patients with Obsessive-Compulsive Disorder. *Journal of Clinical Psychiatry, 67*, 1735–1740.

87. Vanderijst, L., Hever, F., Buot, A., et al. (2024). Psilocybin-Assisted Therapy for Severe Alcohol Use Disorder: Protocol for a Double-Blind, Randomized, Placebo-Controlled, 7-Month Parallel-Group Phase II Superiority Trial. *BMC Psychiatry, 24*(1), 77. https://doi.org/10.1186/s12888-024-05502-y

88. Bogenschutz, M. P., Forcehimes, A. A., Pommy, J. A., Wilcox, C. E., Barbosa, P. C. & Strassman, R. J. (2015). Psilocybin-Assisted Treatment for Alcohol Dependence: A Proof-of-Concept Study. *Journal of Psychopharmacology, 29*, 289–299.

89. Johnson, M. W., Garcia-Romeu, A., Cosimano, M. P. & Griffiths, R. R. (2014). Pilot Study of the 5-HT2AR Agonist Psilocybin in the Treatment of Tobacco Addiction. *Journal of Psychopharmacology, 28.*

90. Johnson, M. W., Garcia-Romeu, A. & Griffiths, R. R. (2017). Long-Term Follow-Up of Psilocybin-Facilitated Smoking Cessation. *American Journal of Drug and Alcohol Abuse, 43,* 55–60.

91. Krupitsky, E., Burakov, A., Romanova, T., Dunaevsky, I., Strassman, R. & Grinenko, A. (2002). Ketamine Psychotherapy for Heroin Addiction: Immediate Effects and Two-Year Follow-Up. *Journal of Substance Abuse Treatment, 23*(4), 273–283. https://doi.org/10.1016/s0740-5472(02)00275-1

92. Krupitsky, E. M., Burakov, A. M., Dunaevsky, I. V., Romanova, T. N., Slavina, T. Y., & Grinenko, A. Y. (2007). Single Versus Repeated Sessions of Ketamine-Assisted Psychotherapy for People with Heroin Dependence. *Journal of Psychoactive Drugs, 39*(1), 13–19. https://doi.org/10.1080/02791072.2007.10399860

93. Ezquerra-Romano, I., Lawn, W., Krupitsky, E., & Morgan, C. J. A. (2018). Ketamine for the Treatment of Addiction: Evidence and Potential Mechanisms. *Neuropharmacology, 142,* 72–82. https://doi.org/10.1016/j.neuropharm.2018.01.017

94. Mithoefer, M. C., Mithoefer, A. T., Feduccia, A. A., et al. (2018). 3,4-methylenedioxymethamphetamine (MDMA)-Assisted Psychotherapy for Post-Traumatic Stress Disorder in Military Veterans, Firefighters, and Police Officers: A Randomised, Double-blind, Dose-response, Phase 2 Clinical Trial. *Lancet Psychiatry, 5*(6), 486–497. https://doi.org/10.1016/S2215-0366(18)30135-4

95. Veterans Exploring Treatment Solutions. (2024). Veterans Exploring Treatment Solutions. Retrieved from https://vetsolutions.org/

96. Williams, S. (2024). Psychoactive Drug Ibogaine Effectively Treats Traumatic Brain Injury in Special Ops Military Vets. *Stanford Medicine News Center.* https://med.stanford.edu/news/all-news/2024/01/ibogaine-ptsd.html

97. Davis A. K., Xin Y., Sepeda N., Averill L.A. (2023). Open-Label Study of Consecutive Ibogaine and 5-MeO-DMT Assisted-Therapy for Trauma-Exposed Male Special Operations Forces Veterans: Prospective Data from a Clinical Program in Mexico. *Am J Drug Alcohol Abuse. 49*(5): 587-596. doi: 10.1080/00952990.2023.2220874

98. Perez, Z. (2023). First-ever Provision for Psychedelic Studies Included in Defense Bill. *Military Times.* https://www.militarytimes.com/newsletters/2023/12/15/first-ever-provision-for-psychedelic-studies-included-in-defense -bill/

99. 1A. (2024). The Power of Psychedelic Therapy for Members of the Military. NPR [Audio podcast]. Retrieved from https://www.npr.org/2024/02/07/1198909987/1a-draft-02-07-2024

100. World Health Organization. (2022). Mental Disorders. *World Health Organization.* Retrieved from https://www.who.int/news-room/fact-sheets/detail/mental-disorders

101. Martinez-Ales, G., Hernandez-Calle, D., Khauli, N., & Keyes, K. M. (2020). Why Are Suicide Rates Increasing in the United States? Towards a Multilevel Reimagination of Suicide Prevention. *Current Topics in Behavioral Neurosciences,* 46, 1–23. https://doi.org/10.1007/7854_2020_158

102. Watts, A. (2013). *The Joyous Cosmology: Adventures in the Chemistry of Consciousness.* New World Library. 27.

103. Wittmann, M. (2023). *Altered States of Consciousness: Experiences Out of Time and Self.* MIT Press. 108–113.

104. Rodríguez Arce, J. M. & Winkelman, M. J. (2021). Psychedelics, Sociality, and Human Evolution. *Frontiers in Psychology,* 12, 729425. https://doi.org/10.3389/fpsyg.2021.729425

105. Pollan, M. (2019). *How to Change Your Mind: What the New Science of Psychedelics Teaches Us About Consciousness, Dying, Addiction, Depression, and Transcendence.* Penguin Press. 288.

106. Huxley, A. (2009). *The Perennial Philosophy: An Interpretation of the Great Mystics, East and West.* Harper Perennial Modern Classics.

107. Frecska, E., Bokor, P., & Winkelman, M. (2016). The Therapeutic Potentials of Ayahuasca: Possible Effects against Various Diseases of Civilization. *Frontiers in Pharmacology,* 7, 35. https://doi.org/10.3389/fphar.2016.00035

108. Gonzalez, D., Cantillo, J., Perez, I., Carvalho, M., Aronovich, A., Farre, M., Feilding, A., Obiols, J. E. & Bouso, J. C. (2021). The Shipibo Ceremonial Use of Ayahuasca to Promote Well-Being: An Observational Study. *Frontiers in Pharmacology,* 12, 623923. https://doi.org/10.3389/fphar.2021.623923

109. Oikarinen, I. (2020). *Forest Medicines: Case Study of the Understandings of Health and Elements of Healing Among Practitioners of Amazonian Yawanawá*

Shamanism (master's thesis). University of Helsinki. Retrieved from https://helda.helsinki.fi/handle/10138/316949

110. Pérez-Gil, L. (2001). The Yawanáwa Medical System and its Specialists: Healing, Power, and Shamanic Initiation. *Cadernos de Saúde Pública, 17,* 333–344. https://doi.org/10.1590/S0102-311X2001000200008

111. Hofmann, Albert., Wasson, G. R., Ruck, C., & Staples, B. (1998). *The Road to Eleusis: Unveiling the Secret of the Mysteries.* Hermes Press.

112. Ruck, Carl. P. & Webster, P. (2006). Symposium: The Mythology and Chemistry of the Eleusinian Mysteries. *Proceedings of the 2006 World Psychedelic Forum conference: LSD.* Switzerland.

113. Muraresku, B. C. (2023). *The Immortality Key: The Secret History of the Religion with No Name.* St. Martin's Griffin.

114. Allegro, J. M. (2009). *The Sacred Mushroom and the Cross: A Study of the Nature and Origins of Christianity Within the Fertility Cults of the Ancient Near East.* Doubleday.

115. Richards, W. (2015). *Sacred Knowledge: Psychedelics and Religious Experiences.* Columbia University Press. 43.

116. Richards, W. (2015). *Sacred Knowledge: Psychedelics and Religious Experiences.* Columbia University Press. 28.

117. Letheby, C. (2021). *Philosophy of Psychedelics.* Oxford University Press. 197–204.

118. Letheby, C. (2021). *Philosophy of Psychedelics.* Oxford University Press. 197.

119. Letheby, C. (2021). *Philosophy of Psychedelics.* Oxford University Press. 202.

120. Pollan, M. (2019). *How to Change Your Mind: What the New Science of Psychedelics Teaches Us About Consciousness, Dying, Addiction, Depression, and Transcendence.* Penguin Press.

Chapter 8: A Call for Democratic Transhumanism

1. Dodes, R. (2019). Bryan Johnson's Antiaging Quest Has Made Headlines. But There's More to His Story. *Vanity Fair.* https://www.vanityfair.com/news/biohacker-antiaging-lawsuits

2. Dodes, R. (2019). Bryan Johnson's Antiaging Quest Has Made Headlines. But There's More to His Story. *Vanity Fair.* https://www.vanityfair.com/news/biohacker-antiaging-lawsuits

3. Alter, C. (2021). The Man Who Thinks He Can Live Forever. *Time*. https:/
    /time.com/6315607/bryan-johnsons-quest-for-immortality/

4. Vance, A. (2023). The Man Who Spends $2 Million a Year to Look 18 Is
    Swapping Blood with His Father and Son. *Bloomberg Businessweek Tech-
    nology*.https://www.bloomberg.com/news/articles/2023-05-22/bryan-johnson
    -s-anti-aging-blood-transfusion-involves-dad-and-son?srnd=premium
    &leadSource=uverify%20wall&sref=MTy2GeXk

5. Jackson, S. (2023). The Tech Exec Who Eats Exactly 1,977 Calories a Day as
    He Tries to Age Backward Says He Has No 'Desire' for Cheat Days—and It
    Makes Him 'Sick' to Think of Pizza and Donuts. *Business Insider*. https://www
    .businessinsider.com/millionaire-tech-exec-bryan-johnson-feels-sick-thinking
    -pizza-donuts-2023-7

6. Mikhail, A. (2023). Who Is Bryan Johnson's Doctor, Oliver Zolman? *Fortune*.
    https://fortune.com/well/2023/07/30/who-is-bryan-johnsons-doctor-oliver
    -zolman/

7. Barrabi, T. (2023). Tech Tycoon Bryan Johnson Touts Penis Rejuvenation
    Therapy. *NY Post*. https://nypost.com/2023/08/17/tech-tycoon-bryan-johnson
    -touts-penis-rejuvenation-therapy/

8. Bostrom, N. (2005). Ethics of Artificial Intelligence and Robotics. In F. Adams.
    (Ed). *Ethical issues for the Twenty-First Century*, (pp. 3-14). Philosophy Docu-
    mentation Center.

9. Huxley, J. (2015). Transhumanism. *Ethics in Progress*, 6(1), 12–16.

10. Clark, A. (2003). *Natural Born Cyborgs: Minds, Technologies, and the Future of
    Human Intelligence*. Oxford University Press.

11. de Grey, A. D. N. J. (2023). The Divide-and-Conquer Approach to Delaying
    Age-Related Functional Decline: Where Are We Now? *Rejuvenation Research*,
    26(3), 217–220. http://doi.org/10.1089/rej.2023.0057

12. Sandberg, A. (2013). Morphological Freedom—Why We Not Just Want It, but
    Need It. In M. More & N. Vita-More (Eds.), *The Transhumanist Reader: Clas-
    sical and Contemporary Essays on the Science, Technology, and Philosophy of the
    Human Future* (pp. 56–64). Wiley-Blackwell.

13. Bostrom, N. (2002). Existential Risks: Analyzing Human Extinction Scenarios
    and Related Hazards. *Journal of Evolution and Technology, 9*.

14. Koene, R. (2013). Uploading to Substrate-Independent Minds. In M. More & N. Vita-More (Eds.), *The Transhumanist Reader: Classical and Contemporary Essays on the Science, Technology, and Philosophy of the Human Future* (pp. 146–156). Wiley-Blackwell.

15. Sandberg, A. (2013). An Overview of Models of Technological Singularity. In M. More & N. Vita-More (Eds.), *The Transhumanist Reader: Classical and Contemporary Essays on the Science, Technology, and Philosophy of the Human Future* (pp. 376–394). Wiley-Blackwell.

16. Tegmark, M. (2017). *Life 3.0: Being Human in the Age of Artificial Intelligence.* Vintage Books.

17. Kendal, Evie. (2022). "Ectogenesis and the Ethics of New Reproductive Technologies for Space Exploration. In E. Tumilty & M. Battle-Fisher (Eds.), *Transhumanism: Entering an Era of Bodyhacking and Radical Human Modification.* Cham: Springer 222.

18. More, M. (1990). The Principles of Extropy. *Extropy: The Journal of Transhumanist Thought, 1*(2), 6–16.

19. Vallor, S. (2016). *Technology and the Virtues: A Philosophical Guide to a Future Worth Wanting.* Oxford University Press.

20. Gebru, T. & Torres, É. P. (2024). The TESCREAL Bundle: Eugenics and the Promise of Utopia Through Artificial General Intelligence. *First Monday, 29*(4). https://doi.org/10.5210/fm.v29i4.13636

21. Market Research Future (MRFR). (2023). Biohacking Market Size to Hit USD 63 Billion by 2028 at 19% CAGR. *Market Research Future (MRFR).* Retrieved from https://www.globenewswire.com/news-release/2023/01/24/2594446/0/en/Biohacking-Market-Size-to-Hit-USD-63-Billion-by-2028-at-19-CAGR-Report-by-Market-Research-Future-MRFR.html

22. Grazia. (2023). The Man Who Biohacked His Way to Billions: Andrew Masanto. *Grazia Magazine.* Retrieved from https://graziamagazine.com/us/articles/the-man-who-biohacked-his-way-to-billions-andrew-masanto/

23. Grazia. (2023). The Man Who Biohacked His Way to Billions: Andrew Masanto. *Grazia Magazine.* Retrieved from https://graziamagazine.com/us/articles/the-man-who-biohacked-his-way-to-billions-andrew-masanto/

24. Rafiq, Q., Christie, L. & Morgan, H. M. (2023). *Biohacking: A Thematic Analysis of Tweets to Better Understand How Biohackers Conceptualise Their Practices.* Retrieved from https://www.medrxiv.org/content/10.1101/2023.02.16.23286022v1

25. Seim, D. (2022). The Exploding Growth of Wearable Medical Devices. *Halcousa*. https://www.halcousa.com/the-exploding-growth-of-wearable-medical -devices/

26. Walker, M. (2017). *Why We Sleep*. Simon & Schuster.

27. TheSleepDoctor. (2024). YouTube. Retrieved from https://www.youtube .com/@TheSleepDoctor

28. Grazia. (2023). The Man Who Biohacked His Way to Billions: Andrew Masanto. *Grazia Magazine*. Retrieved from https://graziamagazine.com/us /articles/the-man-who-biohacked-his-way-to-billions-andrew-masanto/

29. Ferriss, T. (2009). *The 4-Hour Workweek: Escape 9-5, Live Anywhere, and Join the New Rich*. Harmony Publishing.

30. Thompson, C. (2018). *The Science of Fasting: How to Lose Weight, Heal Your Body, and Experience Limitless Energy Now*. Independently published.

31. Atkins, R. C. (1981). *Dr. Atkins' Diet Revolution*. Bantam Books.

32. Asprey, D. (2015). *The Bulletproof Diet: Lose Up to a Pound a Day, Reclaim Energy and Focus, Upgrade Your Life*. Rodale Books.

33. Hyman, M. (2024). *Dr. Mark Hyman*. Retrieved from https://drhyman.com/

34. Greenfield, B. (2024). The Ultimate Guide to Nootropics, Smart Drugs, and Psychedelics. *Ben Greenfield Fitness*. Retrieved from https:/ /bengreenfieldlife.com/article/brain-articles/ultimate-guide-nootropics -smart-drugs-psychedelics/

35. Thesis. (2024). *Personalized nootropics for every brain*. Retrieved from https:// takethesis.com/

36. Wang, L., Wang, N., Zhang, W. et al. (2022). Therapeutic Peptides: Current Applications and Future Directions. *Sig Transduct Target Ther, 7*(48).

37. Wim Hof Method. (2024). Retrieved from https://www.wimhofmethod.com/

38. Greenfield, B. (2023). The Ultimate Guide to Cold Thermogenesis: Health Benefits & How to Do It. *Ben Greenfield Fitness*. Retrieved from https:/ /bengreenfieldlife.com/article/ultimate-guide-cold-thermogenesis-health -benefits/

39. Patrick, R. (Guest), & Rose, K. (Host). (2016). Dr. Rhonda Patrick, Cold Stress and Longevity Hacking [Audio podcast episode]. In the Kevin Rose Show. Apple Podcasts. https://podcasts.apple.com/us/podcast/dr

-rhonda-patrick-on-cold-stress-and-longevity-hacking/id1088864895
?i=1000363894493

40. Grazia. (2023). The Man Who Biohacked His Way to Billions: Andrew Masanto. *Grazia Magazine.* Retrieved from https://graziamagazine.com/us
/articles/the-man-who-biohacked-his-way-to-billions-andrew-masanto/

41. de Grey, A. (2006). A Roadmap to End Aging. TED. Retrieved from https:/
/www.ted.com/speakers/aubrey_de_grey

42. Barja, G. (2014). "The Mitochondrial Free Radical Theory of Aging." In H. D.
Osiewacz (Ed.), *Progress in Molecular Biology and Translational Science* (vol.
127, pp. 1–27). Academic Press.

43. Bowles, N. (2018). Aaron Traywick, Prominent Biohacker, Dies at 28. *New York
Times.* Retrieved from https://www.nytimes.com/2018/05/19/style/biohacker
-death-aaron-traywick.html

44. Zhang, S. (2018). The Death of a Biohacker. *The Atlantic.* Retrieved from
https://www.theatlantic.com/science/archive/2018/05/aaron-traywick
-death-ascendance-biomedical/559745/

45. Earle, J. (2022). Embodiment Diffracted: Queering and Cripping Morphological
Freedom. In E. Tumilty & M. Battle-Fisher (Eds.), *Transhumanism: Entering an
Era of Bodyhacking and Radical Human Modification* (pp. 149–173). Springer.

46. Xu, C., Xie, Y., Zhong, T., Liang, S., Guan, H., Long, Z., Cao, H., Xing, L., Xue, X.
& Zhan, Y. (2022). A Self-Powered Wearable Brain-Machine-Interface System
for Real-Time Monitoring and Regulating Body Temperature. *Nanoscale, 34,*
12483–12490.

47. BGI. (2024). *BGI Genomics.* Retrieved from https://www.bgi.com/us/

48. Wang, L., Zuo, X., Ouyang, Z., Qiao, P. & Wang, F. (2021). A Systematic Review of Antiaging Effects of 23 Traditional Chinese Medicines. *Evidence-Based
Complementary and Alternative Medicine,* 5591573. https://doi.org/10
.1155/2021/5591573

49. Shen, C. Y., Jiang, J. G., Yang, L., Wang, D. W., & Zhu, W. (2016). Anti-Ageing
Active Ingredients from Herbs and Nutraceuticals Used in Traditional Chinese Medicine: Pharmacological Mechanisms and Implications for Drug
Discovery. *British Journal of Pharmacology, 173*(8), 1418–1435. https://doi
.org/10.1111/bph.13631

50. European Union. (2024). *Guangzhou Regenerative Medicine and Health-Guang-dong Laboratory, Centre of Chemistry and Chemical Biology.* Retrieved from https://ai-dd.eu/node/32

51. BGI. (2024). *BGI Genomics.* Retrieved from https://www.bgi.com/us/

52. Raposo, V. L. (2019). The First Chinese Edited Babies: A Leap of Faith in Science. *JBRA Assisted Reproduction, 23*(3), 197–199. https://doi.org/10.5935/1518-0557.20190042.

53. Regalado, A. (2019). China's CRISPR Babies: Read Exclusive Excerpts from the Unseen Original Research. *MIT Technology Review,* https://www.technologyreview.com/2019/12/03/131752/chinas-crispr-babies-read-exclusive-excerpts-he-jiankui-paper/.

54. Sataline, S., & Sample, I. (2018). Scientist in China Defends Human Embryo Gene-Editing. *The Guardian.* https://www.theguardian.com/science/2018/nov/28/scientist-in-china-defends-human-embryo-gene-editing.

55. Perrin, D. , & Burgio, G. (2019). China's Failed Gene-Edited Baby Experiment Proves We're Not Ready for Human Embryo Modification. *The Conversation,* https://theconversation.com/chinas-failed-gene-edited-baby-experiment-proves-were-not-ready-for-human-embryo-modification-128454.

56. Kennedy, M.. (2019). Chinese Researcher Who Created Gene-Edited Babies Sentenced to 3 Years in Prison. NPR. https://www.npr.org/2019/12/30/792340177/chinese-researcher-who-created-gene-edited-babies-sentenced-to-3-years-in-prison.

57. Mark, M. (2024). Elon Musk's Neuralink Brain Implant to Ultimately Help Humans Merge with AI is Sparking Debate over Safety and Ethics. *Business Insider.* Retrieved from https://www.businessinsider.com/neuralink

58. de Grey, A., & Rae, M. (2007). *Ending Aging: The Rejuvenation Breakthroughs that Could Reverse Human Aging in Our Lifetime.* St. Martin's Press.

59. Auer, R.M. (2022). Mens Humana in Corpore Humano—Body-Hacking the Human Experience. In E. Tumilty & M. Battle-Fisher (Eds.), *Transhumanism: Entering an Era of Bodyhacking and Radical Human Modification* (pp. 80–98). Springer. 80.

Chapter 9: Racing to the Cosmos

1. Azarian, B. (2022). *The Romance of Reality: How the Universe Organizes Itself to Create Life, Consciousness, and Cosmic Complexity.* BenBella Books. 279.

2. Sorvino, C. (2022, September 22). California's Water Emergency: Satisfying the Thirst of Almonds While the Wells of the People That Harvest Them Run Dry. *Forbes.* Retrieved from https://www.forbes.com/sites/chloesorvino/2022/09/22/california-farms-pump-water-to-feed-crops-amid-extreme-heat-and-drought-but-residents-wells-are-running-dry/?sh=10f2de706eac

3. Kuzma, S. (2023). Highest Water-Stressed Countries. *World Resources Institute.* Retrieved from https://www.wri.org/insights/highest-water-stressed-countries

4. Hand, K. P., Chyba, C. F., Priscu, J. C., Carlson, R. W., & Nealson, K. H. (n.d.). Astrobiology and the Potential for Life on Europa. *Europa,* 589–629. Retrieved from https://nexsci.caltech.edu/workshop/2019/25_Hand_4020.pdf

5. NASA. (2024). Europa. *NASA Science.* Retrieved from https://science.nasa.gov/jupiter/moons/europa/facts/

6. Barucci, M. A., Fulchignoni, M. & Lazzarin, M. (1996). Water Ice in Primitive Asteroids? *Planetary and Space Science, 44*(9), 1047–1049. https://doi.org/10.1016/0032-0633(96)00002-5

7. Argo Space. (2024). Retrieved from https://www.argospace.com/

8. Roan, A. J. (2021). Moon Mining Challenge Opens New Doors. *Metal Tech News.* Retrieved from https://www.metaltechnews.com/story/2021/06/30/mining-tech/moon-mining-challenge-opens-new-doors/603.html

9. Masten Space Systems. (2024). Retrieved from https://masten.aero/

10. Liquifer Systems Group. (2024). LUWEX: Water on the Moon. Retrieved from https://liquifer.com/luwex-water-on-the-moon/

11. Kurzweil, R. (2006). *The Singularity Is Near: When Humans Transcend Biology.* Penguin Books. 250.

12. Space Solar Ltd. (2024). Retrieved from https://www.spacesolar.co.uk/

13. Orbital Composites. (2024). Orbital Manufacturing. Retrieved from https://www.orbitalcomposites.com/orbital-manufacturing

14. Peters, A. (2024). This Startup Plans to Beam Solar Power from Space. *Fast Company.* Retrieved from https://www.fastcompany.com/91021022/this-startup-plans-to-beam-solar-power-from-space

15. Hurler, K. (2023). Companies Beam Space-Based Solar Power to Earth. *Gizmodo*. Retrieved from https://gizmodo.com/companies-beam-space-based-solar-power-earth-1850561102

16. Fung, B. (2023). The Big Bottleneck for AI: A Shortage of Powerful Chips. *CNN*. Retrieved from https://www.cnn.com/2023/08/06/tech/ai-chips-supply-chain/index.html

17. Toews, R. (2023). The Geopolitics of AI Chips will Define the Future of AI. *Forbes*. Retrieved from https://www.forbes.com/sites/robtoews/2023/05/07/the-geopolitics-of-ai-chips-will-define-the-future-of-ai/?sh=72a77ead5c5c

18. Seligman, L. (2022). China Dominates the Rare Earths Market. This U.S. Mine Is Trying to Change That. *Politico*. Retrieved from https://www.politico.com/news/magazine/2022/12/14/rare-earth-mines-00071102

19. Turton, S. (2024). Australia Funds Rare Earth Research as West Seeks China Alternatives. *Nikkei Asia*. Retrieved from https://asia.nikkei.com/Business/Markets/Commodities/Australia-funds-rare-earth-research-as-West-seeks-China-alternatives

20. Lorinc, J. (2023). Space Mining's Best Prospect Is VC Money, Not Asteroid Gold. *Bloomberg*. Retrieved from https://www.bloomberg.com/news/articles/2023-12-06/space-mining-will-likely-draw-more-venture-capital-money-than-asteroid-metals?utm_source=website&utm_medium=share&utm_campaign=copy

21. Asteroid Mining Corporation. (2024). Roadmap. Retrieved from https://www.asteroidminingcorporation.co.uk/roadmap

22. Astroforge. (2024). Retrieved from https://www.astroforge.io/

23. TransAstra. (2024). Retrieved from https://transastra.com/

24. Williams, M. (2019). Is It Worth It? The Costs and Benefits of Space Exploration. *Interesting Engineering*. Retrieved from https://interestingengineering.com/science/is-it-worth-it-the-costs-and-benefits-of-space-exploration

25. Williams, M. (2021). Why We Need to Keep Going to Space and Shouldn't Fix Earth first. *Interesting Engineering*. Retrieved from https://interestingengineering.com/innovation/why-we-need-to-keep-going-to-space-and-shouldnt-fix-earth-first

26. Chow, D. (2011). Everyday Tech from Space: Out of NASA Tragedy, Better Fireproof Clothes. *Space.com*. Retrieved from https://www.space.com/10671-space-spinoff-technology-fireproof-clothing.html

27. Shelhamer, M., Bloomberg, J., LeBlanc, A., Prisk, G. K., Sibonga, J., Smith, S. M., Zwart, S. R. & Norsk, P. (2020). Selected Discoveries from Human Research in Space That Are Relevant to Human Health on Earth. *NPJ Microgravity, 6*, 5. https://doi.org/10.1038/s41526-020-0095-y

28. Dunbar, B. (2022). Nichelle Nichols Helped NASA Break Boundaries on Earth and in Space. *NASA.* Retrieved from https://www.nasa.gov/people-of-nasa/diversity-at-nasa/nichelle-nichols-helped-nasa-break-boundaries-on-earth-and-in-space/

29. NASA. (2020). Tracy Drain, Systems Engineer. *NASA.* Retrieved from https://www.nasa.gov/image-article/tracy-drain-systems-engineer/

30. Maddox, T. (2020). Tech Leaders Share How Star Trek Iinspired Them to Pursue a Career in Technology. *TechRepublic.* Retrieved from https://www.techrepublic.com/article/tech-leaders-share-how-star-trek-inspired-them-to-pursue-a-career-in-technology/

31. Vallor, S. (2016). *Technology and the Virtues: A Philosophical Guide to a Future Worth Wanting.* Oxford University Press.

32. Kurzweil, R. (2006). *The Singularity Is Near: When Humans Transcend Biology.* Penguin Books. 357–359.

33. Tegmark, M. (2017). *Life 3.0: Being Human in the Age of Artificial Intelligence.* Vintage Books. 241.

34. Bostrom, N. (2008, April 22). Where Are They? *MIT Technology Review.* Retrieved from https://www.technologyreview.com/2008/04/22/220999/where-are-they/

35. Bostrom, N. (2008, April 22). Where Are They? *MIT Technology Review.* Retrieved from https://www.technologyreview.com/2008/04/22/220999/where-are-they/

36. Azarian, B. (2022). *The Romance of Reality: How the Universe Organizes Itself to Create Life, Consciousness, and Cosmic Complexity.* BenBella Books. 133.

37. Azarian, B. (2022). *The Romance of Reality: How the Universe Organizes Itself to Create Life, Consciousness, and Cosmic Complexity.* BenBella Books. 262–264.

38. Lockheed Martin. (2024). Autonomous unmanned systems. Retrieved from https://www.lockheedmartin.com/en-us/capabilities/autonomous-unmanned-systems.html

39. Airbus. (2024). Artificial Intelligence. Retrieved from https://www.airbus.com/en/innovation/industry-4-0/artificial-intelligence

40. Valizadegan, H., Martinho, M. J. S., Wilkens, L. S., Jenkins, J. M., Smith, J. C., et al. (2022). ExoMiner: A Highly Accurate and Explainable Deep Learning Classifier that Validates 301 New Exoplanets. *Astrophysical Journal, 926*(2), 120. https://doi.org/10.3847/1538-4357/ac4399

41. Axiom Space. (2024). Retrieved from https://www.axiomspace.com/

42. NASA. (2021). NASA's Self-Driving Perseverance Mars Rover Takes the Wheel. *NASA*. Retrieved from https://www.nasa.gov/solar-system/nasas-self-driving-perseverance-mars-rover-takes-the-wheel

43. Wodecki, B. (2022). AI to Help NASA's James Webb Telescope Map the Stars. *AI Business*. Retrieved from https://aibusiness.com/verticals/ai-to-help-nasa-s-james-webb-telescope-map-the-stars

44. Sarkar, D. J. (2019). AI-Powered Search for Extra-Terrestrial Intelligence: Analyzing Radio Telescopic Data. *Towards Data Science*. Retrieved from https://towardsdatascience.com/ai-powered-search-for-extra-terrestrial-intelligence-analyzing-radio-telescopic-data-c9e46741041

45. Smith, J. K. (2020). Osteoclasts and Microgravity. *Life, 10*, 207.

46. Williams, D., Kuipers, A., Mukai, C., & Thirsk, R. (2009). Acclimation During Space Flight: Effects on Human Physiology. *CMAJ, 180*(13), 1317–1323. https://doi.org/10.1503%2Fcmaj.090628

47. Fogtman, A., Baatout, S., Baselet, B., et al. (2023). Towards Sustainable Human Space Exploration—Priorities for Radiation Research to Quantify and Mitigate Radiation Risks. *Microgravity, 9*(1), 8. https://doi.org/10.1038/s41526-023-00262-7

48. Arone, A., Ivaldi, T., Loganovsky, K., Palermo, S., Parra, E., Flamini, W., & Marazziti, D. (2021). The Burden of Space Exploration on the Mental Health of Astronauts: A Narrative Review. *Clinical Neuropsychiatry, 18*(5), 237–246.

49. LanzaTech. (2024). Retrieved from https://lanzatech.com/

50. Rampelotto, P. H. (2013). Extremophiles and Extreme Environments. *Life (Basel), 3*(3), 482–485.

51. Ott, E., Kawaguchi, Y., Kölbl, D., et al. (2020). Molecular Repertoire of Deinococcus Radiodurans After 1 Year of Exposure Outside the International

Space Station Within the Tanpopo Mission. *Microbiome, 8,* 150. https://doi.org/10.1186/s40168-020-00927-5

52. Aiyer, K. (2023). How Microbes Could Aid the Search for Extra-Terrestrial Life. *American Society for Microbiology.* Retrieved from https://asm.org/articles/2023/september/how-microbes-could-aid-the-search-for-extra-terres

53. Steigerwald, W. (2018). Mars Terraforming Not Possible Using Present-Day Technology. *NASA.* Retrieved from https://www.nasa.gov/news-release/mars-terraforming-not-possible-using-present-day-technology/

54. NASA Goddard Space Flight Center. (2018). Terraforming the Martian Atmosphere. *NASA Mars Exploration Program.* Retrieved from https://mars.nasa.gov/resources/21974/terraforming-the-martian-atmosphere/

55. Norwitz, N. G., Mota, A. S., Misra, M., & Ackerman, K. E. (2019, March 26). LRP5, Bone Density, and Mechanical Stress: A Case Report and Literature Review. *Frontiers in Endocrinology (Lausanne), 10,* 184.

56. Kendal, Evie. (2022). Ectogenesis and the Ethics of New Reproductive Technologies for Space Exploration. In E. Tumilty & M. Battle-Fisher (Eds.), *Transhumanism: Entering an Era of Bodyhacking and Radical Human Modification,* (pp. 211-226). Springer, 222.

57. Kendal, Evie. (2022). Ectogenesis and the Ethics of New Reproductive Technologies for Space Exploration. In E. Tumilty & M. Battle-Fisher (Eds.), *Transhumanism: Entering an Era of Bodyhacking and Radical Human Modification,* (pp. 211-226). Springer, 222.

Chapter 10: It's All Up to Us

1. Azarian, B. (2022). *The Romance of Reality: How the Universe Organizes Itself to Create Life, Consciousness, and Cosmic Complexity.* BenBella Books. 277.

2. Azarian, B. (2022). *The Romance of Reality: How the Universe Organizes Itself to Create Life, Consciousness, and Cosmic Complexity.* BenBella Books. 277.

ACKNOWLEDGMENTS

In the grand pursuit of self-improvement, there's one essential fuel: knowledge. Just as a fire needs constant tending to burn bright, our intellectual growth thrives on the consistent consumption of information and ideas. Self-improvement is not just about accumulating facts but about cultivating a curious mind that actively seeks out diverse perspectives and challenges existing assumptions. Whether it's delving into the wisdom of ancient philosophers or grappling with the groundbreaking ideas of contemporary thinkers, the act of engaging with knowledge expands our understanding of ourselves and the world around us.

Knowledge and curiosity always grow when a person is surrounded by family that values learning, research, and wild thinking. My deepest gratitude goes to my family for their unwavering support of my . . . well, let's say my unique pursuits: writer, PKM-enthusiast, and explorer of psychedelic-assisted therapy. They've stood by me even when friends and followers questioned my writing or entrepreneurial ventures. As the only one in my immediate family venturing into psychedelic-assisted therapy, I could have faced judgment or fear from them. But they trusted my meticulous research and intentional approach, knowing that I always gather all the knowledge I'll need before trying something new. For that space to think freely, the resources, and the freedom to experiment and learn, I owe them everything.

I'd be remiss, as a young writer, if I did not thank the many fellow thinkers and writers whose work was essential to my own thought process. This journey of intellectual exploration requires a deep appreciation for the tireless efforts of those who contribute to the vast ocean of knowledge. A heartfelt thank-you goes out to the brilliant writers who dedicate themselves to crafting words that illuminate, inspire, and challenge. Their tireless efforts ensure that a constant stream of fresh ideas flows into the public discourse, enriching our understanding of the human experience.

A debt of gratitude is owed to Dr. Ben Kuipers for his willingness to share his expertise and insights on the issues of trust and cooperation. My extended interviews with Dr. Kuipers provided invaluable depth and perspective to this exploration of achieving moonshot moments. My lighthearted and enjoyable conversations with Ben shed light on the crucial role cooperative principles play in collective human advancement. Dr. Kuipers's contribution has been instrumental in shaping a more nuanced understanding of how we can foster a society that thrives on collaboration and shared goals.

A heartfelt thank-you goes out to father and son team Dr. Bill Richards and Dr. Brian Richards for their extraordinary generosity and contributions to this work. Our interviews provided profound insight into the development of the field of psychedelic research and the potential opportunities for improved mental health treatments in the future. Their willingness to open their homes and share their wealth of knowledge on psychedelics was truly humbling. Beyond the meaningful interviews, they provided invaluable resources and reading materials, fostering a deeper understanding of the field. Engaging with such titans of psychedelic research was a true honor. Their dedication to pushing the boundaries of this transformative field is an inspiration, and their openness in sharing their expertise has enriched this book immeasurably.

I want to thank Angela for being the psychedelic-assisted therapist I needed. You helped me rediscover my strength, my capacity for softness, and how to reconnect with my humanity, soul, and sense of self. I'm grateful for the space you always made for me, the trust you placed in me, and for guiding me toward the colors when the world seemed black and white. You empowered me to become the Stoic I always wanted to be, not by suppressing my emotions or traumas, but by teaching me to observe them without judgment and reconcile them. I'm forever grateful for the tools you've given me to live a life less burdened by my own thoughts.

Along the way, I must also extend my gratitude to the courageous researchers, mystics, therapists, and shamans in the field of psychedelics. Their unwavering dedication to pushing the boundaries of scientific exploration has kept the field alive despite decades of government interference, social taboo, and propaganda. Their work holds immense promise for the future of mental health, and their commitment to rigorous research paves the way for responsible therapeutic applications.

Furthermore, a deep bow goes out to the innovators who are building companies with a socially beneficial purpose. These visionaries understand that success is not solely measured in profit margins but also in the positive impact a company can have on society. Their dedication to addressing real-world problems and working for the public good serves as a beacon of hope, demonstrating the power of innovation to create a better future for all.

I'd be remiss not to express my sincere gratitude to my publisher Health Communications, Inc., and distributor Simon & Schuster for their support of this book. A special thanks goes out to Greg Brown, whose editorial expertise and insightful commentary proved instrumental in refining these ideas and bringing them to life. His keen insights and meticulous attention to detail helped shape this manuscript into a stronger, more cohesive work.

Much like describing a psychedelic experience, words seem inadequate to express my gratitude to Dr. M. Chloe Mulderig for her contributions to everything I do. Her blend of artistic curiosity and mental discipline is unmatched; she helps me organize my thoughts and vision better than any human or AI I've encountered. When I'm too bogged down to have wit, Chloe provides it. When discipline wanes, she steps in with unwavering focus. It's as if my favorite philosophers' references to transformative mentors come alive in Chloe. Our many debates about religion, history, ethics, and cultural evolution over the years have proved essential to how I view faith and belief throughout history. Her constant presence, wisdom, and articulation have profoundly shaped my internal dialogue for the better. In simpler terms, if I were Batman, Chloe would be my Lucius Fox—there's no dream too big she wouldn't tackle to make it a reality. Thank you for making this book and its ideas accessible to the world and for always making me seem like I have it all together.

In that spirit, the finest seats at my intellectual table are always reserved for the philosophers. These tireless thinkers grapple with the big questions—the nature of reality, the meaning of life, the foundation of morality. Their explorations, even when controversial, push us to think critically about ourselves and the world we inhabit. Their pursuit of wisdom serves as an inspiration to all who seek a deeper understanding of the human condition.

By acknowledging the vital role these individuals play in the ongoing pursuit of knowledge, we all can reaffirm our commitment to continuous learning. As we embark on our own self-improvement journeys, let us celebrate the collective effort that fuels our intellectual growth. This spirit of gratitude will undoubtedly fuel our appetite for knowledge and propel us further on the path of self-improvement.

ABOUT THE AUTHOR

Milan Kordestani is a Gen Z thought leader, bestselling author, and entrepreneur examining the interplay between technology, ethics, and society to advance human cognition, well-being, and cooperation. As the founder of Ankord Media and Ankord Labs, he combines his expertise in design, product development, and startup building with a network of advisors to support impactful companies with strategic investment, intuitive product experiences, and exceptional branding. His *Wall Street Journal* bestseller, *I'm Just Saying,* offers practical tools for bridging divides in a polarized world. His YouTube series, NextPlay, delves into cutting-edge ideas like biohacking, the transformative potential of psychedelic therapy, and the ethical considerations of transhumanism through engaging interviews with leading experts. Deeply passionate about the intersection of technology and humanity, Kordestani emphasizes advancing well-being and global cooperation in an evolving world.